Contents

Swets & Zeitlinger Publishers
Lisse, Abingdon, Exton (PA), Tokyo

intellect™
EXETER, ENGLAND

Paperback Edition First Published in 1998 by
Intellect Books, School of Art and Design, Earl Richards Road North, Exeter EX2 6AS

ISBN 1-871516-62-5

A catalogue record for this book is available from the British Library

Cloth Edition First Published in 1998 by
Swets and Zeitlinger Publishers
Heereweg 347B, 2161 CA LISSE, The Netherlands

ISBN 90-265-1524-3

Library of Congress Cataloging-in-Publication Data

IT for learning enhancement / edited by Moira Monteith.
p. cm.
Includes bibliographical references and index.
ISBN 9026515243
1. Computer-assisted instruction--Great Britain. 2. Information technology--Great Britain.
I. Monteith, Moira.
LB1028.43.I816 1998
371.33'5--dc21
98-6283
CIP

Consulting Editor: Masoud Yazdani
Copy Editor: Lucy Kind
Cover Design: Sam Robinson
Production: Richard Monteith

Printed and bound in Great Britain by Cromwell Press, Wiltshire

Contributors

Chris Abbott lectures in the School of Education at King's College, London where he is also involved in research with young people, language and the Internet. He has a particular interest in the use of the World Wide Web by special schools, marginalised groups and young people. He has worked at national level on the development of ICT educational policy in South Africa and in several European countries. He writes widely on aspects of language and ICT, both in academic publications and in The Times Educational Supplement.

Lyn Dawes has been a primary science teacher for many years, during which time she worked with an Open University research team on the SLANT and TRAC projects, looking at children's talk as they used computers. She is now a research student and lecturer at De Montfort University, Bedford, studying the impact of the introduction of the National Grid for Learning on the professional development of teachers.

Toni Downes is the Associate Dean of the Faculty of Education at the University of Western Sydney, Australia. Her teaching responsibilities include educational computing subjects in the undergraduate and postgraduate courses. Her research interests include children's use of information technologies in information handling and communication processes and more generally for learning across the curriculum. Her current research project explores the impact of children's home access to computers on schooling. She is co-author of two recent books: *In Control: Young Children Learning with Computers* and *Learning in the Electronic World.*

Helen Finlayson has come from a background of science, psychology and education and has been working with children and teachers using ICT over a number of years. **Deirdre Cook**, a senior lecturer in education, has a particular interest in literacy and other early semiotic activity in young children. Recently they have worked together productively, with other colleagues, on developing ICT across the curriculum, the use of ICT for collaborative learning and on the role of laptops in emergent literacy. Helen Finlayson lectures at the University of Warwick, Institute of Education and Deirdre Cook at the University of Derby, School of Education and Social Sciences.

Carol Fine is Subject Leader for ICT at the School of Education at Kingston University. She recently moved from the University of North London where she worked with Mary Lou Thornbury in establishing ICT programmes for teacher training and for INSET. In addition to her interest in the contribution of computer control to children's learning in the early years she is working with Mary Lou Thornbury to follow up the use of ICT by teachers in their first teaching post. Their study focuses on how ICT courses in teacher training contribute to the use of ICT by newly-qualified teachers.

Jonathan Grove is studying for his Ph.D in the School of Cultural Studies, Sheffield Hallam University. He has been working with pupils in a number of Sheffield schools, observing their responses while using a virtual environment he has constructed using the VR authoring package Superscape VRT. His most recent field study involved 30 children, working in groups, examining and commenting on a Greek villa he designed. He is particularly interested in the use of VR as a stimulus for discussion by pupils and students. He has delivered a number of conference papers both in the UK and the USA.

Stephen Marcus, Graduate School of Education, University of California, Santa Barbara, co-ordinates technology-related projects for the National Writing Project, including those featuring telecommunication, video, laptops, photography, and teacher-research. He has served on the Commission on Media for the National Council of Teachers of English and is currently on its Committee on Information Literacy and the Executive Board for its Assembly on Computers and English. He has received awards from the International Society for Technology and Education, Computer-Using Educators, and the Alliance for Computers and Writing.

Moira Monteith lectures in the School of Education, Sheffield Hallam University, predominantly on inservice and postgraduate courses. She has worked on projects with young children and their use of computers for several years. She is currently interested in the spread of telecommunications such as computer conferencing for students, teachers and pupils. With colleagues, she is engaged in linking all partnership schools with the University and, through Sheffield LEA, all Sheffield schools. She is a joint tutor on international computer conferencing courses linking Sheffield Hallam students with students abroad.

Bill O'Neill lectures at the University of Ulster at Coleraine. His academic background is in psychology and he has wide experience of teaching in primary schools in the UK and abroad. He has researched with Harry McMahon for over a decade, looking at the effects of ICT on the literacy of young children. Currently his work involves the training of primary school teachers, with special focus on language and literacy.

Mary Lou Thornbury is ICT Co-ordinator at the School of Education at the University of North London. Over the last five years she has worked with Carol Fine establishing ICT programmes for teacher training and for INSET. In addition to her interest in the contribution of computer control to children's learning in the early years, she is working with Carol Fine to follow up the use of ICT by teachers in their first teaching post. Their study focuses on how ICT courses in teacher training contribute to the uses of ICT by newly-qualified teachers.

Dr Jean Underwood currently holds the post of Reader in Education at the University of Leicester. She has two key roles within the School of Education: as Deputy Director for Research and as Co-ordinator for Information and Communication Technologies . A former chair of the Association of Information Technology in Teacher Edcuation, she is now co-editor of the international journal *Computers & Education*. Her research interests are concerned with supporting children's cognitive and social development through the use of information and communication technologies.

Rupert Wegerif is a Research Fellow in the Centre for Language and Communications in the School of Education at the Open University. He has written a number of articles, chapters and reports on the use of ICT to support learning through communication. His research focuses on the relationship between new media and conversational reasoning. Rupert is editor, with Peter Scrimshaw, of the book *Computers and Talk in the Primary Classroom* published by Multilingual Matters.

Noel Williams is Reader in the Communications Department, School of Critical Studies, Sheffield Hallam University. He is the author of many books in the field of communication and computing. His teaching responsibilities include both postgraduate and undergraduate courses in Communications. His recent research interests include the application of a wide range of computer applications to learning.

Introduction

Moira Monteith

It is perhaps not accidental that politicians in several countries began to call for a 'back to basics' campaign in education around the same time that computers were becoming familiar objects in the classroom. Even more significantly, computers were also available to compile and organise educational statistics. The 'basics' issue has centred mainly on traditional approaches to learning, particularly as regards literacy and numeracy, with the implication that such approaches offer better value to both teachers and students. However, it is computers that have delivered sufficient power to make a centralised curricular system easier to organise and monitor. So whatever happens in the technological future (which is notoriously difficult to predict) the societal impact of computers on the education system cannot be underestimated.

Additionally, in the UK we seem to have moved very rapidly from a time when the education system was blamed for a variety of social ills, to the present when colossal benefits are expected to flow from education. Accompanying the new faith in schools is a belief in the benign power of information communications technology (ICT). In the foreword to *Connecting the Learning Society*, the Government's Consultation Paper on the National Grid for Learning (1997), the Prime Minister, Tony Blair states:

> we intend to lift educational standards in Britain to the level of the best in the world. This will mean making the most of technological change. Technology has revolutionised the way we work and is now set to transform education......Standards, literacy, numeracy, subject knowledge - all will be enhanced by the Grid and the support it will give to our programme for school improvement.

As something of a technophile myself, I hope he's right. Arguably, educational computing could break the confines of the classroom. There is even a possibility that 'official' learning could bypass schools. Or that schools will become very different from the way in which they are currently organised. Learning, even as viewed in the traditional sense of 'school work', can now clearly take place in the home. Awareness of these factors is widespread. Already the Netherlands Education System is considering introducing the concept of 'study rooms' or 'study houses' rather than classrooms in secondary education. A recent report from the Netherlands Government states that 'education has lost its monopoly with regard to learning processes.' They suggest an approach centred on individual needs and motivation 'when a workload made to measure is the most effective.' To achieve this, they also expect an expansion in the use of educational technology[1]. Schools in many countries may become officially places which promote opportunities for social mixing. Drama, sport and other group activities may dominate the school curriculum, but learning the basics (whatever we define them as) may be done at home or in a library or at some site other than school.

I know that many people are worried that in education, movements towards centralisation on one hand and the individual pursuit of learning and training on the other, built on the use of ICT, appear to conflict. Centralised curricula are now a matter of course with all the accompanying machinery of standards regulation, inspection, national targets and publication of league tables projected from a range of data. Moves towards greater use of technology and the beginning of the 'Information Age' should, and maybe will, lead to a reduction of multi-tiered hierarchical structures to perhaps a three tier structure where the customer/client/pupil/student (all consumers) will be more closely in touch with the upper provider/enabler tier than they have ever been before. Students will have the chance to plan their own routes through compulsory education and afterwards, moving easily from one institution to another.

Recent conversations with many colleagues in various educational contexts have revealed much anxiety. Is ICT freedom-giving or imprisoning? How should we react if the answer is yes on both counts? Has the expenditure on ICT in schools proved to be cost-effective so far? Quite possibly not, but schools will be expected to spend a greater proportion of their income on ICT in the future, according to *Connecting the Learning Society.* Will we be able, personally, to manage the explosion of information? Will people be in contact with the world and sit mainly in one room with a computer? If ICT in education brings new skills will it not also de-skill? On a very practical level in the UK, and to quote Professor Stephen Heppel (1997):

> How can we manage with the curriculum nailed down and technology bursting through the roof?

In the recently published document, *Information Technology in English Schools,* on the inspection findings 1995-6, Gabriel Goldstein writes about 'strategic issues':

> The challenge for a large proportion of schools, especially those serving areas where the ownership of home computers is low, is to enhance opportunities for their pupils to use IT by substantially improving their provision. This, of course, is not a comfortable message at a time when the technology itself is changing rapidly.

ICT and Learning Enhancement will, I trust, go some way towards dispelling discomfort. The contributors all have a positive technological stance. The book reveals how alluring some features of ICT can be, the motivation and sense of creativity it can bring. However, the use and implementation of ICT in learning needs to be carefully contextualised with due note taken of research findings and good practice where learning gains have been substantiated. General progress in learning can be built only on the findings of research intelligently addressed. If technology is to 'transform education' as Mr Blair proposes, then we as teachers and learners need to pay attention to evidence from appropriate research. Even more importantly, given the likely speed of change, we need to continue reflecting and evaluating as we implement the use of ICT in classrooms, libraries and homes.

The chapters are all based on research but are not organised to one format alone. Some are more orientated to formal research practice, others to reflection on observed practice in the classroom or at home, while others follow an action research model. At several points they overlap; that is, the focus of attention and the findings inter-relate from chapter to chapter. It is impossible therefore to find one sequence of chapters that fully reveals all the interconnections. I am entirely sure you will be able to find even more links between the chapters than I can. So please accept the list of contents as a provisional sequence only and read them, as no doubt you will, in the sequence you find most appropriate.

Rupert Wegerif and Lyn Dawes begin with a common situation in primary classrooms - a group of children sitting in front of a computer, working together. A previous study, the Spoken Language and New Technology (SLANT) project, revealed that the children's talk was often disappointing from an educational point of view because of problems within the group or because the children did not know how to develop discussion. Other studies confirmed that the quality of interaction is extremely important. Researchers then worked closely with teachers to develop a double strategy. Teachers and children participated in a programme designed to encourage listening skills, develop co-operation but also to stimulate critical argument. This educationally-effective talk has been designated as 'exploratory talk'. Towards the end of this 'intervention programme' children used computer software which encouraged discussion while at the same time allowing them to work independently of the teacher.

Two examples of software designed to support exploratory talk include a science simulation of plant growth, showing how this is affected by temperature, light and water plus a 'talk-support module'. Secondly, a branching narrative geared towards discussion of citizenship helps children talk about moral issues concerned with 'telling tales'. A comparison between two classes revealed that the target class increased its scores on a reasoning test significantly more than the control class did. Transcripts of two groups in the target class revealed that the children spent much more time discussing options and decisions than those in the control class. The talk of both target groups could be categorised as 'exploratory'. Subsequent work has not revealed that such an intervention programme improves learning in more content-based curriculum areas. Therefore the research continues, investigating new ways of supporting exploratory talk and the possibility that children need to be encouraged to recall and share together what they already know in a particular area before moving forward in their work.

Jean Underwood explores those factors that hinder or help effective group work. She looks at research findings as to why group work may or may not be beneficial for children's cognitive as well as social development, and also considers the practical problems facing teachers when they try to encourage group work in school. Her very thorough analysis helps us sift through the many findings concerned with the outcomes of working in groups.

She considers that 'many difficulties are not fundamental weaknesses but arise because insufficient attention is paid to teaching children to work in groups.' She claims however, that evidence shows that children tend to work more collaboratively when working on computer tasks together rather than other classroom tasks. She focuses on her own research while working with a number of colleagues and examines gender differences, finding that girls tend to be more successful at working collaboratively but that when boys are working together and have been instructed that their co-operative work with a partner is most valued, they had the largest performance gains. She quotes evidence revealing that the nature of the outcome measures can have a significant impact on the results of such studies. She looks at a variety of tasks including reading interactive books (a multimedia reading activity) and concludes that 'even in tasks with limited cognitive load, working together can have beneficial effects.' Gender, group composition and the nature of the task can mean that children working in groups have significantly different experiences.

Stephen Marcus writes about an action research project to enrich the participants' understanding of themselves as professionals and their understanding of their students. It was grounded in the notion of the 'reflective practitioner' and ambitiously combined writing with the development of visual literacy. He defines visual information literacy as including the ability to 'read' (ie interpret, decode, translate) and design (ie create and communicate with) images. He believes that there is a reciprocal relationship between teachers studying their teaching and the evolving conduct of that teaching. 'Teachers who observe, question and analyse their own classrooms find that such classroom-based research benefits themselves and their students.' He gives an example of a teacher working with aphasic patients and experienced in the traditional forms of didactic therapy. 'I ask questions to which the patient knows I already know the answers. It is not a substitute for communication.' Photo journals taken by patients 'shifted the format of the spouse's conversation from one of primarily questions to one of comments.' The chapter includes two short but powerfully reflective assignments from teachers.

Stephen Marcus calls photography both 'low tech' and 'old technology,' but overall he reveals the significance it still has for interpreting our own experience - and that of others, as the photographs taken by the aphasic patient showed. The combination of photography and writing or talking are not what we now consider usually multimedia to be but their close connection here signifies learning outcomes not possible from the use of one medium alone.

Toni Downes has been researching extensively the use of computers in Australian homes. She begins by reporting on a young girl's use which is wide, varied and frequent at home but rare in the classroom. The lack of school use is partly because the computer is not often used anyway and when it is, it tends to be used by boys.

Statistics about Australian use of home computers indicate the same factor as in the UK, that there is a definite gap between technology rich homes and those which are technologically poor. Toni Downes comments on the family rules established about the use of the computer. In general, the rules favoured the use of the computer for work or school-related activity over leisure. Males and females, either adult or child, were divided in their choice of games according to gender. Many children considered that they learnt useful skills from playing computer games and the concept of 'play' was very important in their computer use. Much activity was concerned with school work, with which parents helped considerably. Many children were learning 'simple maintenance and problem solving skills' relating to the actual use of the computer. However if something went wrong with the computer, usually the children called a male relative rather than a female.

The majority of children preferred to work on the computer at home rather than at school and were often indignant about the lack of organisational control at school over regulating equal access. A teacher's programme for utilising computer work in the classroom with her children is included. The chapter ends with suggested recommendations as to the better educational use of ICT and the nurturing of links between school and home. Parents were positive about their children's use of computers, believing that they were important for future prospects, but were concerned about handwriting, spelling, reading and using a library. The children do not appear to share those concerns. Toni Downes suggests there is a problem for the children who are losing the motivation to be 'bi-literate', that is combining new literacy skills with the old.

Moira Monteith has been analysing the conversation between children and parents while they were word processing at home. The various factors in the learning context are examined. For instance, there are government pressures to encourage schools to make links with parents and to set more homework. In the near future, documents will be drawn up to define parents' and schools' responsibilities for the children's education. The learning context includes schools' models of learning practice and the parents' own experiences of learning.

Transcripts of the talk are compared with David Wood's model of teacher talk. The dialogue is examined within four groupings: two children from Y1 who are beginning to read and write, four children from the same school who have extremely good 'collections' of printed work, two boys who are not as keen as their mothers would like them to be, and children who are working by themselves. The children had considerable control over the word processing, although parents maintain a very watchful eye over spelling, punctuation and general layout. All tapes show examples of co-operative effort, with some parents going further in the negotiation of a 'learning relationship' than others. Parents who used a question/answer approach had a similar effect to teachers who used that strategy: short answers with no elaboration. Fixed observation of the screen, so well described by Helen Finlayson and Deirdre Cook in their chapter, appeared to lead to excessive comment on spelling and punctuation errors. Mostly the children took such comments in their stride or even ignored them. The transcripts gave ample evidence that many children were developing independent learning skills.

Chris Abbott begins his chapter with two interesting definitions of the World Wide Web. He has been researching into the uses that young people make of this medium for their homepages. He compares this opportunity for publication with more traditional outlets, and claims that there is little in common. He accepts that his chapter may be more in the way of 'historical understanding' rather than 'illumination of current practice', as the home-page culture will undoubtedly have 'moved on' during the comparatively long drawn-out procedure of publishing this book. He makes the analogy that the first web pages were often like brochures, just as the first CD Roms were often versions of books already published.

He contacted a number of young authors of web pages in 1996, mainly male and half of them in North America. He claims that class and race seem less important characteristics in web communication, particularly where pseudonyms are used. In terms of age, many young people are far more expert than older web authors who often seek the advice of younger creators. He suggests that the expertise of young people could be put to good use in schools, where secondary students could help pupils in primary schools design and put up their own web pages.

By 1997 the WWW offered schools a publishing base 'for truly multimedia texts' at surprisingly low costs but many schools have not understood the opportunities available. Chris Abbott believes that we are in a transition stage 'on the way to widespread internet access in the home'. This publication phenomenon could be the impetus for a change in classroom practice. Poetry, stories and opinions can all be made available to audiences of a size and scope never previously envisaged. New virtual communities of young people are developing and this needs to be recognised by schools. He proposes ways in which the curriculum and teacher practice should develop and transform to meet this requirement.

Helen Finlayson and Deirdre Cook have undertaken work with young children to investigate the contribution of the computer itself to their learning rather than the computer and the software together. They distinguish between active and passive software and include a useful diagram of the relationship between active, passive, closed and open software. Passive software could be defined as those programs that reflect the users' ideas back to them for their consideration, and in such a way can aid the creative thinking process. They compared children's responses when offered two tasks, as similarly constructed as possible, but with one presentation on screen and the other off screen. *My World* software, which they considered 'an example of good passive software,' was selected for the research project. Evidence was gathered which revealed that when the children were working with the computer their sessions were longer, they were more involved with their partner's work, less open to general distractions and showed greater persistence and enthusiasm compared with when they were working off the computer. The researchers also suggest that the notion of closure might be important in computer tasks. They found some gender differences in the ways boys used the keyboard but not when they were using the mouse. As part of their summary, they consider that a teacher has a very important role to play in setting appropriate tasks and matching the level of challenge to the children's abilities.

'Children in control' indicates by its name what it is about: children taking the opportunity to advance their independent learning. It also happens to be about 'control', a shorthand term which freakishly lends itself to jokes as to who takes control etc These feeble jokes perhaps reveal the fears that many primary teachers and students have, concerning this area of ICT. It seems so much more technical than databases or spreadsheets and consequently, in apprehension, more difficult than other more 'practical' areas such as painting or textile work. MaryLou Thornbury and Carol Fine not only show us how straightforward it is to give children experience with robots but also how significant it can be in children's cognitive development. They consider whether the introduction of control can promote new ways of learning and thinking and whether children using control can show evidence of the high order thinking and learning that they have not always been able to demonstrate within the traditional curriculum. Their account of how significant control can be for young children is very convincing. It is tightly connected with children's ideas of sequence in narrative as well as investigation and experimentation. 'Control is a creative activity requiring the technological approach of trial, error, evaluation and testing.' At each stage of the work presented here they consider the connections with views of children's development and thinking as proposed by Vygotsky and Papert. Since this is generally viewed as a 'difficult' area for both students and teachers Carol Fine and MaryLouThornbury give outlines for 'progression in controlling and monitoring' from years 1 to 6 to enable others to profit from their vision and experience.

Bill O'Neill argues that computer based multimedia provides educationalists with a powerful set of tools. We will have some influence on the evolution of these tools, since our everyday use will affect that evolution. Educationalists are positioned at a fulcrum where they can both reflect the changing cultural values and exploit technological developments. He traces evolutionary changes through speech and oral traditions, literacy and numeracy, printing and finally, the computer. He too maintains that the teacher is the key figure to the success of computer initiatives in education.

A decade of classroom research by Bill O'Neill and Harry McMahon give them the evidence to argue that learning environments can be created to enhance good classroom practice. Children need to own their work and their learning. By using a digital camera, children 'know that they own the images'. They can then use them 'multifariously', for example in multimedia autobiographies. He considers how language use affects learning and explains how an apprenticeship model encourages children to work collaboratively. He advocates eloquently the case for using photography just as Stephen Marcus does. The chapter concludes that 'Teaching is a deliberate and conscious activity which seeks to help the learner to direct his/her own thinking in a conscious and deliberate manner and to develop an increasing ability to represent his/her own thinking.'

In stimulating contrast, Noel Williams looks at educational multimedia and examines the concept of 'interaction' and to what extent most software claiming to be interactive actually is. He notes that the student is offered more responsibility as well as more choice and needs to know what are the best choices for his/her needs.

Arguably the learner interacts with the designer of the learning system and the system is seldom an 'information resource' but a 'learning model'. The learners' interactions are therefore limited by the designers' model of learning. Indeed it is the information content which is the real benefit of multimedia systems not interaction. The most prevalent multimedia systems follow an instructional model of education.

We, as teachers, are unlikely to be able to predict the actual skills needed with future technology. Such technologies as the use of scanners in supermarkets, interrogating BT information systems or using voice mail are widespread before they get to schools. Pupils and students need real educational value alongside ICT skills. He considers the impact of interactive games and concludes that primary interaction for many players may not be with the game but around the game. 'Computer games set up expectations for interaction which educational multimedia often does not satisfy.' Cost also limits interactivity.

He concludes that hypermedia allows everything to be linked to everything just like human imagination. But teaching, as opposed to learning requires rather narrower perspectives. We need more radical reviews of what learning might be, combined with innovation in design to realise more fully the true potential for educational interaction in multimedia.

The chapter 'Explorations in Virtual History', discusses some of the preliminary findings from work on Virtual Reality (VR) in schools and the implications for future development of this research. The study is concerned with the emergence of a new technological medium and therefore prone, as it must be, to the problems of using such a medium. However, Jonathan Grove and Noel

Williams skilfully outline their research and its relationship with other educational media. They begin by stating the difficulties arising from the fact that VR is an umbrella term for a 'complex network of ideas'. However, the two authors look at the definition as it corresponds to educational use. They outline the case-study, focusing in particular on the role of exploratory talk as it was evidenced in transcripts. They point out also that suggestions such as 'click on that' are indeed quite different in a 'conceptualised three dimensional space' and much more positive than merely a suggestion of a place where the mouse controller might place the pointer. It is therefore perhaps never entirely appropriate to comment in text without video examples when dealing with something such as VR and does lead to the obvious notion that such analysis would benefit from being published on web pages. The case study also provides evidence that assuming that VR is 'intuitive' may be misleading. Currently VR is 'only a little like reality' and therefore children may need to learn to use the medium just as they learn to use any other medium and they need to be 'literate in its codes.' The authors end by considering a set of principles for working practice and indicate how VR might fit in with or be adapted for learning within the classroom or at home.

Note

1 I wish to thank M. Snoek and D Wielenga for the information they have given me concerning changes to the education system in the Netherlands

Encouraging Exploratory Talk Around Computers

Rupert Wegerif and Lyn Dawes

Introduction

Computers in primary schools are mainly used by more than one child at a time. This appears as an efficient use of a relatively scarce resource. Teachers, when asked, also frequently justify the use of group work at computers as a support for peer learning and the development of communication skills (Crook, 1994). The idea that computers can provide a good focus for collaborative learning and communication is supported by a number of studies (Howe, Tolmie, and Mackenzie, 1996; Hoyles, Healy, and Pozzi, 1994; Light, 1993; Littleton, 1995). However when the Spoken Language and New Technology (SLANT) project team video-taped children in ordinary classrooms using computers together, teachers were disappointed by what they saw (Mercer, 1994; Dawes, Fisher and Mercer, 1992). In most cases the children observed were not collaborating effectively at all and showed little sign of learning from each other. In this chapter we will begin by describing this problem, the ineffective way in which children often interact around the computer, then we will describe how we developed an educational intervention to help overcome it, and finally we will present a part of the evaluation of the effectiveness of the programme we have developed. The evidence we present is taken both from a completed study (Wegerif, 1995, 1996) and from an ongoing research project (Wegerif, Mercer, Littleton and Dawes, in press)

The 'base-level' quality of talk around the computer

1) Evidence from the SLANT project

The Spoken Language and New Technology (SLANT) project looked at how primary school children in 12 schools in south east England talked and worked together at the computer. Over 50 hours of video-recordings of classroom activities were collected, transcribed and analysed (Fisher, 1992, 1993; Mercer, 1992, 1994b; Wegerif and Scrimshaw, 1997). On the whole the educational quality of the talk observed was disappointing. Here is a list of some of the strategies used by the children and the problems which emerged from the study:

- one person appointed themselves leader, sitting centrally to the keyboard, and reading from the screen. They called out instantaneous responses to questions, and keyed them in. Other members of the group would agree, or start a futile 'Yes it is/No it isn't' exchange
- children with home computers would become impatient with others who had no keyboard skills, and would again dominate both the keyboard and the decision-taking. Alternatively, a quiet but literate child would work as secretary to a dictator
- less confident children would watch, agree, or withdraw, contributing little. If things subsequently seemed to go wrong, they were castigated for 'not helping'
- friends at work together simply agreed with one another. Other children always disagreed with whatever was suggested, but offered no alternatives
- the content of the talk was observed to be directed towards a re-establishment of the children's friendship groups, or otherwise
- the most heated discussions were to do with who was seated where, who pressed the next key, and so on. Children spent a lot of time talking about how to make the task of actually operating the computer 'fair', an impossibility, but of great importance to them
- talk became general and relaxed if the computer was sited out of the teacher's natural range. This was possibly because children realised that concentrating on the work would mean that their long-awaited turn at the computer would be over sooner and so they chatted about other things
- children competed within the group, using the computer program as a game of some sort. Useless disputes ensued without a constructive outcome.

2) Evidence from a more experimental study

As part of the evaluation of this study we undertook to compare systematically the talk between children in a target class, a class of 9 and 10-year-old children which had worked through our intervention programme, with the talk between children in a control class of the same age in the same school but who had not worked through our intervention programme. For the purposes of this comparison we used a single item of software dealing with citizenship issues called 'Kate's Choice'

(Figure 1). This software had been specially designed to support discussion. However even with this software children who had not been taught how to work together reproduced the same or similar behaviour to the children observed in the SLANT project. Here we will present two illustrations taken from the two 'focal-groups' in the control class who were video-taped. In setting up the exercise the teacher asked the groups to talk together about the questions presented by the software. When we join them the children are faced with a moral dilemma posed by the main character in the story, Kate. (Later in this chapter we will return to this same decision point to describe how children responded after our intervention programme).

Figure 1. A discussion point in the Citizenship curriculum software

Group 1. Mary, Cathy, Brian

Cathy reads from the screen (Figure 1)

Mary: Doesn't tell or tells? What should we do? Does she tell or doesn't she?

Cathy: We've got to guess.

Brian: Tells (in a loud and authoritative voice).

Mary: Tells (Clicks).

Cathy: (Reads from screen) 'Do you all agree?'

Brian: Yes (again in a loud and authoritative voice).

Cathy: Yes.

Total time on the decision screen: 48 seconds.

Commentary

The group do not know what to do despite the cue on the screen to 'Talk about what Kate should do' and the teacher's prompt before they use the software suggesting that they should talk together. Cathy says 'We've got to guess' implying perhaps that she thinks that there is a right answer and that they have to just guess which of the two it is. The one boy in the group decides for everyone with a single authoritative exclamation. No reason is given for his decision. No one questions it.

Group 2. Jim, Tony, Susan

All three read screen together (Figure 1)

Jim: (Reads) 'Talk about what Kate should do then click on one of the buttons.'

Tony: What should we do?

Jim: Do that.

Tony: (Turning to call the teacher) Excuse me. (Turning back to others) We don't know what to do.

Susan: (Clicks)

Jim: Yes we do.

Susan: (Reads) 'Do you all agree?'. Should we tell?

All: Yes.

Total time on the card: 62 seconds.

Commentary

Here again the children are uncertain as to what to do. The cue 'Talk about what Kate should do' means nothing to them. Although Jim reads it out and says that they should just 'do that' he does not in fact try to discuss the issues. Susan, in control of the mouse, takes the decision for the group by clicking it without asking anyone or getting agreement. Nobody protests.

These two 'focal groups' were selected by the class teacher as being representative of the class as a whole. They were the only groups we video-taped. However we gave the same software to the whole control class and took notes on their strategies. We found that most groups moved forward through the story in similar ways to the two above. They exhibited three main strategies:

- unilateral action by the child with the mouse
- accepting the choice of the most dominant child without supporting reasons
- drifting together to one or other choice without debating any alternatives.

The evidence gathered from this study, when combined with the findings from the SLANT project, helps us to establish a 'base-level' account of the quality of 'normal' interaction of children working together at the computer without any prior intervention. Later in this chapter we will describe how children responded to exactly the same computer prompt after our intervention programme and we will offer a quantitative comparison of the talk between groups from the target class and groups from the control class using this same software. But before we do that we will explain how we developed our intervention programme dealing first with our understanding of a desirable way of interacting around the computer which we are calling exploratory talk.

The nature and significance of exploratory talk

In the previous section we have seen that group work around the computer is not always an effective vehicle for learning. The educational benefits of children working together depend on how they interact and particularly on the way in which they talk together. Underwood (1994) supports this point and offers a case study illustrating what he puts forward as 'the type of discussion to be fostered if a successful collaboration is to be seen in the computer classroom'. This type of discussion is characterised by a constructive relationship in which disagreement is accepted and leads to hypotheses being explored together. In characterising his ideal type of productive discussion Underwood draws attention to the work of Kruger (1993). Kruger recorded and coded the talk of pairs working on socio-moral problems and found that the quality of the outcome was related to the quality of the dialogue, particularly the amount of 'transactive reasoning' described as 'criticisms, explanations, justifications, clarifications and elaborations of ideas'. Kruger argues that it is neither conflict nor co-operation that is important in collaborative learning but a combination of the two in a form of interaction which encourages critical challenges within a co-operative search for the best solution.

A range of studies of collaborative work at computers support Underwood and Kruger suggesting that the quality of students' interactions is crucial to the learning outcome. Blaye *et al.* (1991) report that disagreement in itself is less important than the fact that it stimulates verbalisation. Light *et al.* (1994) conclude from a range of studies of pair work on computer-based problems that the style of interaction is more predictive of post-test gains than initial differences in perspective. They argue that having to use language to make plans explicit, to make decisions and to interpret feedback seems to facilitate problem solving and promote understanding.

On the basis of the data collected by the SLANT project Mercer (1994, 1995) came to roughly the same conclusion as Kruger and, following the work of Barnes and Todd in the seventies (Barnes, 1976; Barnes & Todd, 1978), he used the term exploratory talk, for the educationally desirable type of talk that emerged from this research. Exploratory talk is a style of interaction which combines explicit reasoning through talk involving identifiable hypotheses, challenges and justifications, with a co-operative framework of ground-rules emphasising the shared nature of the activity and the importance of the active participation of all involved. Mercer suggested that exploratory talk was one of a limited number of fundamental 'types of talk' or 'social modes of thinking' found in peer collaborations in classrooms. Three of these types of talk found in the SLANT data were described as follows:

- **Disputational talk**, which is characterised by disagreement and individualised decision making. There are few attempts to pool resources, or to offer constructive criticism of suggestions. Disputational talk also has some characteristic discourse features, short exchanges consisting of assertions and challenges or counter assertions

- **Cumulative talk**, in which speakers build positively but uncritically on what the other has said. Partners use talk to construct a 'common knowledge' by accumulation. Cumulative discourse is characterised by repetitions, confirmations and elaborations

- **Exploratory talk**, in which partners engage critically but constructively with each other's ideas. Statements and suggestions are offered for joint consideration. These may be challenged and counter-challenged, but challenges are justified and alternative hypotheses are offered (cf. Barnes and Todd, 1978). Compared with the other two types, in exploratory talk knowledge is made more publicly accountable and reasoning is more visible in the talk.

(Taken with minor modifications from Mercer (1995)p 104 .)

Wegerif and Mercer (1997) expand on this typology by characterising each type of talk, at four levels of description from the fundamental orientation of participants, through the ground-rules being used in the discussion and the type of speech acts produced, down to the actual words used. In Wegerif and Mercer's account exploratory talk is seen to share some ground-rules with cumulative talk, in that it has the same co-operative nature, while also differing because challenges and criticisms are allowed, even encouraged, and these challenges do not break the solidarity of the group. Combining the evidence from research on collaboration with our theoretical framework we came up with the following list of ground-rules for effective exploratory talk in the classroom context. The first three are concerned with creating and maintaining group solidarity, that exploratory talk shares with cumulative talk, the last four are specific to exploratory talk and are concerned with making the talk not just co-operative but also effective for learning:

- all relevant information is shared
- the group seeks to reach agreement
- the group takes responsibility for decisions
- reasons are expected
- challenges are accepted
- alternatives are discussed before a decision is taken
- all in the group are encouraged to speak by other group members.

An educational programme teaching exploratory talk

Once we have a description of the ideal type of talk in terms of social ground-rules it becomes relatively easy to see how this type of talk can be taught or encouraged. These ground-rules are rules guiding practice rather like the rules of a game. To teach them is to draw children into a social practice through what Rogoff describes as 'guided participation'. Collins, Brown and Newman (1986) describe the three phases of cognitive apprenticeship as follows:

- **modelling**, when teachers make explicit what is required and model it for the learners
- **coaching**, when teachers support students' attempts at doing the task
- **fade out**, when students are empowered to continue independently.

The intervention programme we devised to encourage more exploratory talk combined explicit teaching and modelling of the ground-rules from the front of the class with coaching, where the teacher guides the children in discussion and, finally, with fade out, where children work together on activities while the teacher keeps mostly in the background. Each lesson combines the three components of directive teaching, teacher-led whole class discussion and small group work (in groups of three or four established by the teacher to include mixed ability and mixed gender).

The programme consists of a series of nine lessons each designed to last for about one hour. The first part of this teaching programme is designed to encourage the children to think about the purposes of talk, about how they had learned to talk themselves, and how they could describe different ways of talking. After this, the teaching programme focuses on one or more of the ground-rules in each lesson. The early lessons include exercises teaching sub-components of exploratory talk such as effective listening, giving information explicitly and co-operating as a group. For example in one of these lessons children use construction toys to build a model and then have to describe how to replicate this model to a partner who could not see it. The development of these lessons was influenced by the work of the National Oracy Project (Norman, 1992; Open University, 1993). Later lessons encourage critical argument for and against different cases. In one lesson called Dog's Home, for example, groups of children have to match pictures of six dogs with descriptions of five potential owners and decide which dog will have to be left out. The children are given opportunities to practise discussing alternative ideas, giving and asking for reasons, and ensuring all members of the group are invited to contribute. They are told specifically that a major aim of their work as a group is to improve the quality of their talk. The aim of better talk is constantly mentioned, and good practice amongst the groups is encouraged.

Once they have some experience of group work, by lesson three in the current programme, the children are drawn, through a teacher-led discussion, to create and decide upon their own set of ground-rules. The result is then displayed prominently in the classroom and referred to in cases of uncertainty or dispute. Each time the intervention programme has been run these rules are different but nonetheless show considerable overlap with our list of the ground-rules of exploratory talk given above.

These rules give a structure for work around the computer. Work on computer programs is used at the end of the intervention programme as the fade out part of the overall educational strategy. Computer software can support children's discussion while giving them a chance to work independently of the teacher. Decisions about control of the keyboard or mouse, who makes written notes if required, who sat where for how long, and so on, are discussed and sorted out before work actually begins.

Designing computer activities to support exploratory talk

The intervention programme was intended to improve the quality of work at the computer generally by giving the children a model of how to approach working together. However, while the expectations of children working together at the computer is vital to the quality of that work, software factors also have an influence which should not be ignored (Anderson, 1993; Fisher, 1992; Wegerif, 1996; 1997). Much software used in schools has clearly not been designed to support group work or group discussion. The conclusion of the SLANT project was that it is necessary to look at the whole educational activity around the computer if we are to improve the quality of the associated learning (Mercer, 1994). We have responded to this need in two ways. First, we have designed new items of software and software features specifically to encourage exploratory talk. Secondly, we have designed classroom activities around existing software, particularly CD-Roms, in a way intended to encourage exploratory talk.

Some principles for designing computer-based activities to support exploratory talk

Analysis of data gathered by the SLANT project produced several principles for both teachers and software designers who want to encourage exploratory talk around computers (Wegerif, 1997):

1 Make sure problems are sufficiently complex

Only some kinds of problems are good for supporting discussion. Problems posed have to be sufficiently complex to benefit from the multiple perspectives brought to bear in discussion. They should also have significant consequences for the users perhaps through the way in which they are embedded in a developing narrative (Phillips and Scrimshaw, 1997).

2 Provide props for reasoning

Props should be provided in the form of supporting information or ready-made arguments for and against different positions. A 'notebook' for recording results so far can also serve as a shared focus for debate. These props might be provided by the software programme itself or through materials made to be used with the programme.

3 Don't let them turn it into a competitive game

It is wise to design software in a way which resists rather than encourages competitive game-playing use. Time measures should be avoided for this reason as should having small discrete tasks towards a single goal which the users could transform into competitive turn-taking.

4 Discourage mental turn-taking

Physical turn-taking might be a way of making sure that everyone in the group is involved but it must not be conflated with the mental turn-taking where users do not discuss what they are doing together but each acts separately taking responsibility for their choices.

5 Encourage role-play and narrative

Children appear to discuss things most freely at most length in the context of being drawn into narratives within which they play roles or discuss the implications of roles.

6 Minimise typed input

Typing appears to be such a daunting task for most children that engaging in it takes over the direction of the whole exercise. Teaching children keyboard skills at an early age might be another approach to overcoming this problem.

Figure 2. A prompting screen encouraging the pupils to discuss explicit predictions

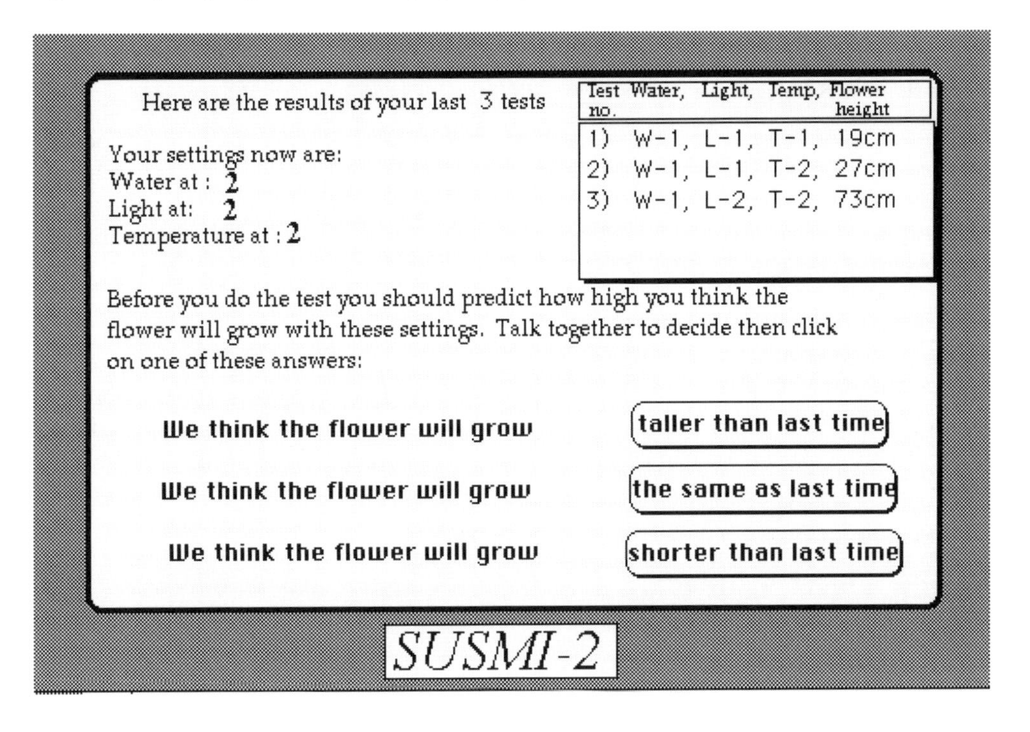

Two examples of software designed to support exploratory talk

A supported science simulation

We designed an item of software based around a simple simulation of plant growth, showing how this is affected by temperature, light and water. This basic design is similar to software simulations commercially available. To support exploratory talk two extra dimensions were added. First, the simulation was embedded in an overall narrative frame in which the children role-played scientists trying to find the formula to help a friend win the local flower-show. Second, what amounted to a 'talk-support' module was added. This module interrupted the users as they tried to run the simulation prompting them to talk to each other, asking them to make explicit predictions, relate these predictions to outcomes and explain why the predictions were either right or wrong.

Figure 3. Near the end of the story the users have to decide collectively whether they made Kate do the right thing or not

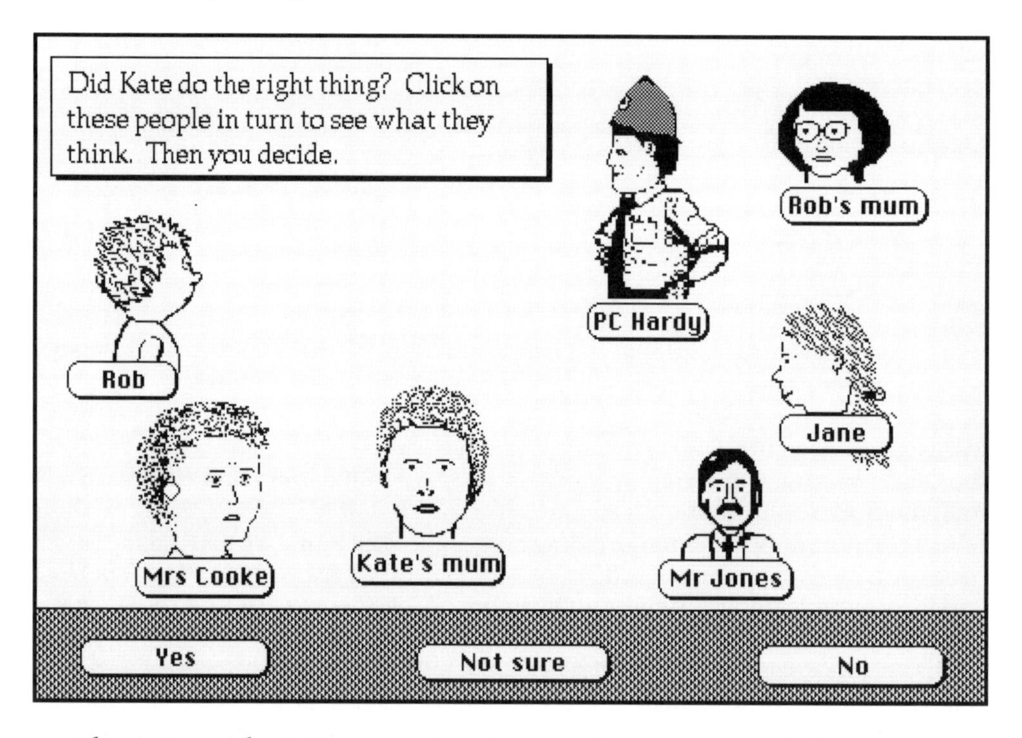

The design of the 'talk-support' module was based on research done at Strathclyde University on computer-based collaborative learning in science (Tolmie & Mackenzie, 1996; Tolmie *et al.*, 1993). This research indicates that significant learning gains follow if children are encouraged first to discuss predictions before conducting experiments and then relate the outcomes to those predictions.

A branching narrative for citizenship

An item of computer software was also designed to integrate with the area of 'citizenship' as an extra activity after whole-class discussion about the issue of stealing from shops. A screen from this program has already been shown in Figure 1. This software takes the form of a branching narrative about a girl called Kate divided between loyalty to her friend and pressure to tell others about his theft. In the course of the story Kate meets the shopkeeper and others involved in the crime and they give their points of view. The group of children using this software have to make decisions as to what Kate should do or say at key junctures in the story, and these decisions determine how the story continues (Figure 3).

The aim is not to direct the children towards a particular conclusion but to encourage genuine and wide-ranging debate about the issues. The theory behind this is that the capacity to 'de-centre' sufficiently to take the point of view of others into account is the core component of moral development (Rowe, 1992; Rowe and Newton, 1994).

Designing activities around CD-Roms

As well as designing educational software to support exploratory talk we have designed educational activities using existing software, particularly CD-Rom encyclopaedias, which can be incorporated into our intervention programme. Evaluation of the use of CD-Roms in schools indicates that there is a need to develop ways of converting the information available on large CD-Rom databases into curriculum knowledge. Too frequently children are drifting in an undirected way through a sea of information (Collins *et al.*, 1995). Discussion is one way of helping children put the information they can access on a CD-Rom into a personal and curriculum context. To this end we have produced worksheets which prompt the children to talk together before they access a CD-Rom encyclopaedia in order to find out what they know already and specify what they are trying to learn. On returning from the CD-Rom they are prompted to note down whether what they have found together fitted or did not fit with what they expected to find. This is a variation on the 'predict, observe, explain' model widely used in science education (Harlen, 1992).

In a current example of this approach, the groups of children were asked to consider four animals, using a CD-Rom information finder. They were given printed information in which each animal was pictured and briefly described. The children were asked to discuss the animals, and decide together whether each was a herbivore, carnivore, or omnivore. They recorded their decision, and, briefly, their reasons for the decision, and then used the CD-Rom to access factual information. Finally they looked at whether their reasons had led them to the correct conclusion.

Evaluating the effectiveness of the intervention programme

The overall effectiveness of the intervention programme in improving the quality of the children's talk together was evaluated by giving them a reasoning task to work on in groups of three both before and after the intervention programme. This task consisted of a set of problems taken from Raven's progressive matrices so it resulted in a quantitative measure of the change in the effectiveness of the group process. A comparison between the target class and a control class using this measure showed that the target class increased its scores on the reasoning task significantly more than the control class over the period of the intervention. Three focal groups in the target class were recorded solving these reasoning problems. Discourse analysis of the change in their talk over the period of the intervention showed that it increased in features associated with exploratory talk such as asking many more task-focused questions and using many more connectors such as 'because' and more hypotheticals such as 'if'. Detailed discourse analysis further showed a clear connection between this shift in the use of language and the increased solving of reasoning problems. More details of these results are given in Wegerif (1996) and in Wegerif and Dawes, (1997).

The change in the way groups worked together around computer software was also assessed. This was done through a combination of video-tapes of focal groups in the target class and in the control class, discourse analysis of transcripts of their talk and automatic data collection. To give a worked example of part of this evaluation we will return to the issue of talk around a decision point in the programme 'Kate's Choice' which is reproduced in Figure 1. The talk of the two focal groups in the control class has already been reported. The control class had had no intervention lessons of any sort. The target class, who had had nine weeks of our special exploratory talk lessons, responded very differently. Here are transcripts of the talk of the focal groups in the target class in response to the same screen in the program:

Group 1. Natalie, Jane and George

Natalie reads the instructions from the screen (Figure 1).

Jane: Right we'll talk about it now.

Natalie: Ssh (reads) 'talk about what Kate should do. When you have decided click on one of the buttons'.

Jane: Well what do you think ?

George: Doesn't tell.

Jane: What do you think Natalie?

Natalie: Well I think she should tell because it's wrong to steal - but it's her friend.

Jane and George (together): It's her friend.

Natalie: It's her friend as well.

George: He knows it's wrong.

Natalie: Yes but he's not doing it for her, er, for him, he's doing it for his mother. So I reckon she doesn't tell.

Jane: Yes, I agree.

George: Agreed, agreed.

Natalie: Doesn't tell then? One, two, three - (clicks).

George: Here we go, here we go.

Jane: (reads) 'Have you all talked about it?'

All: Yes.

Natalie: (Clicks.)

Total time on the card: 97 seconds.

Commentary

These children respond immediately to the cue on the screen which says 'Talk together'. They obviously know what this means and they sit back from the screen a little and turn to look at each other. Jane takes on a discussion facilitator's role asking the others what they think and encouraging a consensus. Through this everyone is involved. Reasons are given and questioned. Both Natalie and George give reasons against their original positions. Natalie appears to change her view. All children reach agreement before the mouse is clicked.

According to the description of exploratory talk put forward earlier this talk is clearly exploratory. Reasons for assertions are given and questioned within a cooperative orientation.

Group 2. Barbara, Martin and Ross

Barbara reads aloud from the screen (Figure 1).

Ross: I think he should not - he shouldn't tell.

Martin: Don't tell.

Ross: (Reads) 'Talk about what Kate should do ...'

Ross: I think she shouldn't tell because she said she'd promise.

Martin: Yeh, if she broke her promise he'd be into trouble right? Broke her promise he'd be into trouble.

Barbara: Yeh, but on the other hand...?

Ross: Yeh and he did do it not for himself but for his mum and his mum's sick.

Martin: No, but he could be lying.

(3 second pause)

Barbara: Yeh, but would you do it? - would you tell?

Ross: Umm, no. If I did I'd feel guilty.

Martin: I wouldn't.

Barbara: (Clicks and then reads) 'Have you talked about it?'

ALL: Yes.

Total time on the card: 82 seconds.

Commentary

Here Barbara takes on a facilitating role asking questions and putting forward alternatives. She challenges the sincerity of the others asking them if they would really do what they are saying Kate should do. Nobody argues in favour of telling but Michael suggests that they should be cautious in believing Robert's story. They all reach agreement before the mouse is clicked. Again the talk is exploratory with children raising and criticising a range of reasons for both alternatives before reaching a shared decision.

The following features were exhibited in the talk of most of the target class groups observed:

- asking each other task-focused questions
- giving reasons for statements and challenges
- considering more than one possible position
- drawing opinions from all in the group
- reaching agreement before acting.

These features were very different from the features of the control class groups described in the second section of this chapter. To counter the possible accusations that our observations were biased and that we had carefully selected the focal groups to give the results we wished, we added a quantitative dimension to the evaluation. This relied on the fact that groups who discussed more took longer at each decision point. The control class and the target class were highly comparable. They were year 6 classes (9 and 10-years-old) of the same size and average ability in the same school. Both classes were given the same software divided into groups of three and asked to talk together. Automatic data collection recorded how long each group took at each screen. The results of the first decision point illustrated in Figure 1 are compared in the following table.

Table I. Time in seconds spent on the first decision point of Kate's Choice for all groups from both target and control classes

	Target class groups	Control class groups
	43	21
	63	35
	65	41
	67	48*
	74	51
	82*	58
	97*	59
	102	60
	105	62*
Mean	77.55	48.33
S D	20.72	13.76

(* = focal group.)
Statistical analysis of these figures shows them to be highly significant ($p = 0.0015$. One-tailed T-test).

This quantitative comparison combined with the earlier qualitative comparison of the talk of the focal groups, indicates that our intervention programme achieved its aim of improving the educational quality of the talk around the computer.

Summary and conclusions

We began this chapter by raising the issue of the quality of children's talk around computers. Research evidence suggested that this was often disappointing from an educational point of view and yet computers are mainly used by groups of children. We went on to describe how we developed an intervention programme to improve the quality of talk around computers. This intervention combined lessons teaching exploratory talk, a kind of talk which both research and conceptual analysis proposed as the ideal kind of talk for collaborative learning, with the development of software to support exploratory talk between groups of children. We tried to show how both aspects of the intervention were based on design principles emerging from our research and that of others. Finally we described a part of the evaluation of the intervention programme which showed that it had had a very positive effect on the educational quality of children's talk around a particular piece of software.

The results we have presented were taken from a project which has already been completed. However we are now working on a larger project implementing this same intervention in several middle schools. As part of this project we are developing further software and computer-based educational activities. The positive results we have gained so far are enough to show that this work is worth pursuing but they do not indicate that we have fully solved the problem yet. We have shown that our intervention programme encouraging exploratory talk can improve group reasoning, both in solving abstract puzzles of the kind given in standard tests of reasoning and in the discussion of socio-moral issues. It has been more difficult for us to show that this intervention improves learning in more content-based curriculum areas. We are exploring the different ground-rules and combination of ground-rules which work best to promote learning in different areas. We are also exploring new ways of supporting exploratory talk and learning through exploratory talk in computer-based activities around stand-alone machines and also through computer networks.

References

Anderson, A. Tolmie, A. McAteer, E. & Demisne, A. 'Software Style and Interaction Around the Microcomputer'. *Computers and Education* **20** (3) (1993): 235-250.

Barnes, D, & Todd, F. *Discussion and Learning in Small Groups*. Routledge & Kegan Paul: London, (1978).

Barnes, D. *From Communication to Curriculum*. Penguin Books: London, (1976).

Blaye, A. Light, P. Joiner, R. & Sheldon, S. 'Collaboration as facilitator of planning and problem solving on a computer-based task'. *British Journal of Developmental Psychology* **9** (1991): 471-483.

Brown, J. S. Collins, A. & Duguid, P.' Situated Cognition and the Culture of Learning'. *Educational Researcher* **1** (1989).

Collins, A. Brown, J. S. & Newman, S. E.' Cognitive Apprenticeship'. In Resnick L. B. (ed) *Cognition and Instruction: Issues and agendas*. Lawrence Erlbaum Associates: Hillsdale, New Jersey, 1986.

Collins, J. Longman, J. Littleton, K. Mercer, N. Scrimshaw, P. & Wegerif R. *CD-Roms in Primary Schools: An independent evaluation*: Centre for Language and Communication, Open University, UK, 1995.

Crook, C. *Computers and the collaborative experience of learning*. Routledge: London and New York, 1994.

Dawes, L. Fisher, E. & Mercer, N. 'The Quality of Talk at the Computer'. *Language and Learning* (1992): 22-25.

Fisher, E. ' Characteristics of children's talk at the computer and its relationship to the computer software'. *Language and Education* **7** (2): (1992) 187-215.

Fisher, E. ' Distinctive features of pupil-pupil talk and their relationship to learning'. *Language and Education* **7** (4): (1993) 239-258.

Harlen, W. ' Research and Development of Science in the Primary School'. *International Journal of Science Education* **14** (1992).

Howe, C. Tolmie, A. & Mackenzie, M. 'Computer support for the collaborative learning of Physics concepts'. In O'Malley C. (ed) *Computer-supported collaborative learning*. Springer-Verlag: Berlin, 1996.

Hoyles, C. Healy, L. & Pozzi, S. 'Groupwork with computers'. *Journal of Computer Assisted Learning* **10** (1994): 202-215.

Kruger, A. 'Peer collaboration: conflict, co-operation or both?' *Social Development* **2** (3): (1993).

Light, P. 'Collaborative learning with computers'. In Scrimshaw, P. (ed) *Language, Classrooms and Computers*. Routledge: London, 1993.

Light, P. Littleton, K. Messer, D. & Joiner R. 'Social and communicative processes in computer-based problem solving'. *European Journal of Psychology of Education*, 1994.

Light, P. and Littleton, K. 'Cognitive approaches to group work'. In Kutnick, P. & Rogers, C. (eds) *Groups in Schools*. Cassell: London, 1994.

Littleton, K. 'Children and computers'. In Bancroft D. & Carr R. (eds) *Influencing development*. Basil Blackwell, 1995.

Mercer, N. 'The quality of talk in children's joint activity at the computer'. *Journal of Computer Assisted Learning* **10** (1994): 24-32.

Mercer, N. *The Guided Construction of Knowledge: talk amongst teachers and learners*. Multilingual Matters: Clevedon, 1995.

Norman, K. (ed) *Thinking Voices: The work of the national oracy project*. Hodder and Stoughton: London, 1992.

Open University. *Talk and Learning 5-16: an in-service pack on oracy for teachers*. The Open University: Milton Keynes, 1991.

Phillips, T. & Scrimshaw P. 'Talk round Adventure Programs'. In Wegerif, R. & Scrimshaw P. (eds) *Computers and Talk in the Primary Classroom*. Multi-lingual Matters: Clevedon, 1997.

Rowe, D. The citizen as a moral agent - the development of a continuous and progressive conflict-based citizenship curriculum. *Curriculum* **13** (3): (1992).

Rowe, D. & Newton J. *You, Me, Us: Citizenship materials for primary schools*. The Citizenship Foundation: London, 1994.

Tolmie, A. Howe, C. Mackenzie, M. & Greer, K. Task design as an influence on dialogue and learning: primary school group work with object flotation. *Social Development* **2** (3): (1993).

Underwood, G. 'Collaboration and problem solving: gender differences and the quality of discussion'. In Underwood J. (ed) *Computer based learning*. David Fulton: London, 1994.

Wegerif, R. *Computers, Talk and Learning: using computers to support reasoning through talk across the curriculum*. Ph.D, Institute of Educational Technology, Open University, Milton Keynes, UK, 1995.

Wegerif, R. 'Collaborative learning and directive software'. *Journal of Computer Assisted Learning* **12** (1) (1996): 22-32.

Wegerif, R. 'Factors Affecting the Quality of Children's Talk at Computers'. In Wegerif, R. & Scrimshaw, P. (eds) *Computers and Talk in the Primary Classroom*. Multi-lingual Matters: Clevedon, 1997.

Wegerif, R. 'Using computers to help coach exploratory talk across the curriculum'. *Computers and Education* **26** (1-3): (1996) 51-60.

Wegerif, R. & Dawes, L. 'Computers and exploratory talk: an intervention study'. In Wegerif, R. & Scrimshaw, P. (eds) *Computers and Talk in the Primary Classroom*. Multi-lingual Matters: Clevedon, 1997.

Wegerif, R, & Mercer, N. 'A dialogical framework for researching peer talk'. In Wegerif, R. & Scrimshaw, P. (eds) *Computers and Talk in the Primary Classroom*. Multi-lingual Matters: Clevedon, 1997.

Making Groups Work

Jean Underwood

Introduction

The arguments promoting group work in schools include the notion that group work provides a supportive and secure learning environment for children, that it promotes 'real' world skills that will be needed in the workplace, that it involves children actively in the learning process; and relationships between peers and between children and the teacher will improve as a result of this (Reid, Forrestal and Cook, 1982). In addition, in multi-cultural settings co-operative activities allow children to gain an understanding and an appreciation of others' perspectives (Cowie, 1994; Slavin, 1993). The rationale supporting group work has to do with the ways in which children learn to think. The argument here is that in any group that effectively collaborates, the members of the group will be able to introduce knowledge and ideas to the other members and will in turn be able to accept information from their partners. However, this view that working in small groups will result in an additive effect, producing cognitive benefits not available to the individual worker (Schoenfeld, 1989) has not always been born out by the research evidence (Hill, 1982).

Turning aside from the value of group work for a moment it should be recognised that the myth that such learning environments are the norm in UK primary schools has been exposed by Galton, Fogelman, Hargreaves, and Cavendish (1991). Their evidence shows that most of the time children work as individuals even though they are physically organised in groups of 4, 6 or 8 to a table. Classroom observations conducted as part of the SCENE project showed that, although 78 per cent of all observed activity in junior classrooms took place in such physical organisations, with a further 10 per cent taking place in pairs, for only 14.5 per cent of the time that children were seated together were they working together. For over 50 per cent of the time children worked individually. The remaining time was time-off-task including general socialisation.

Effective group work does not simply follow from teachers' acceptance of such a model of organisation of the classroom. Indeed many teachers' first experience of co-operative grouping is less than satisfactory and viewed as more trouble than it is worth, hence the retreat to the setting of individual work (Alexander, Willcocks and Kinder, 1989). One reason for this is a general lack of understanding of why groups do and do not work. This chapter will explore those factors that hinder or help effective group work. For sometime now we have been investigating the learning outcomes that take place when children, mainly in pairs, work with and around computers. This work has two defining foci: an emphasis on co-operative and collaborative working, and an emphasis upon predictive models of learning outcomes. Evidence from our own controlled interventions in schools and that of other researchers has provided insight into the social and cognitive circumstances of effective collaborations with computers.

The nature of group work

We have already seen that sitting children in groups does not necessarily mean they are working together; they may be engaged in co-operative activities but not necessarily collaborative learning. Terms such as group work, co-operative and collaborative learning are used very loosely in the literature but here refer to small groups working to achieve a common goal. In achieving that common goal, however, the members of the group may choose to take responsibility for sub-tasks and work co-operatively, or they may collaborate and work together on all parts of the problem (Dunne and Bennett, 1990). The production of a class newspaper is a good example of a co-operative activity. Some children will act as journalists/ writers, others will edit, while others will work on layout and design. All the tasks are important for the production of the newspaper but individuals or small groups of children work largely independently in fulfilling their own part of the assignment. A typical collaborative task would be children brain-storming a solution to a problem. If the learners collaborate and share in the decision making process the level of social interaction is necessarily high, but this is not necessarily so for co-operative workers. 'Collaboration is a co-ordinated, synchronous activity that is the result of a continued attempt to construct and maintain a shared conception of a problem' (Teasley and Roschelle, 1993). It may be that some tasks are more likely to engender collaborative rather than co-operative strategies as we shall show from our own studies, but equally the characteristics of the social group may predispose the members to one mode of learning rather than another.

The case for and against working in groups

Let us return to why group work should be effective. Kruger (1993) suggests that when pairs of children solve a problem together, they think more effectively than when they work alone. Recent studies of the benefits of group work have emphasised the Piagetian idea of socio-cognitive conflict resulting in decentralised thinking or the Vygotskian perspective by which social co-operation delivers new cognitive products. Thus the acquisition of a concept depends on comparing and contrasting, which is most effectively achieved through interaction with peers (Brown and Palincsar, 1986). Teasley and Roschelle (1993) describe this as a sharing of the 'joint problem space', where the participants use language and action to establish shared knowledge, to recognise divergence from that shared knowledge, and to rectify misunderstandings that impede work.

Not all of the evidence supports this view that group work results in improved performance, for example Light and Colbourn (1987) report no advantages for groups in a microPROLOG programming task when compared with the performance of individuals. Schwartz (1995) cogently argues, however, that this failure to show the benefits of group work is a product of the outcome measures used in many studies. He contends that it is not surprising that efficiency measures, such as time to complete a task or the percentage of correct answers, often fail to tap into the cognitive benefits of group work, particularly for short duration studies. This is because it takes time to acquire an initial understanding of the co-worker's point of view and understanding of the task. His results from three problem solving tasks, comparing the performance of individuals and dyads, have shown that time taken in gaining that understanding of another's point of view, is not time wasted. Schwartz provides convincing evidence in support of Brown and Palincsar's assertion that it is through this development of a mutual reference point that children engage in higher cognitive skills that lead to abstract thought and the use of effective representations. In his three problem solving tasks the children working in pairs brought more complex and more complete solutions to the task than a comparable set of children who were working as individuals.

If group work is deemed a sound educational strategy why does so little take place? The reality is that there are many difficulties facing teachers when they try to organise group work. Field research has shown a high incidence of both physical and verbal abuse that renders the groups unproductive (Cowie, Smith, Boulton and Laver, 1994). This research shows that not all group discussion leads to the rich and fruitful collaboration encapsulated in Teasley and Roschelle's sharing of the 'joint problem space'. In many instances discussion may simply express the etiquette of turn-taking (low-level co-operation) or, in dysfunctional groups, may be directed to non-task derogatory comments (non co-operation). The result is that teachers are prone to abandon group work if their initial attempts are unproductive and fail to deliver the deep discussions predicted by many learning theorists (Biott, 1987). However, many of these difficulties are not fundamental weaknesses but arise because insufficient attention is paid to teaching children groupwork. Brown (1988) points out, there is a 'two-stage process in learning to work in a group'. In the first stage, members have to learn how to submerge their 'personal identity' and to acquire a social one. He contends that this initial period where children have to acquire this social identity, can be a very difficult one. Children either use avoidance strategies to evade participating in the group or adopt more aggressive roles as described by Cowie *et al.* (1994).

There may also be task effects on group work. It has been argued that while the social cohesiveness of the group is developing, more complex tasks requiring abstract reasoning should be avoided. In such situations, if a problem is set which can have a variety of answers, pupils will tend to pick the first answer and stick with it. This reduces the possibilities of participation. For this reason, practical tasks that have a specific solution or problems with only one answer are likely to produce more continuous discussion.

The debate on the effectiveness of group work is still open. There is a growing body of evidence to confirm the view that working in groups and with peers is an effective way of achieving some educational goals, particularly those associated with higher order thinking rather than rote learning of facts. There is an equal body of evidence that group work is often ineffectual and that the failure is at least partly due to the social structure of the group, although task variables may also be important. I turn now to our own research that focuses on groups working with computers, to tease out some of the factors that facilitate or inhibit effective interaction and therefore learning.

The child, the group and the computer:- gender issues

Working in a group is a social act but how do variations in the nature of the group affect the nature of the interaction and therefore the learning that takes place? For example, if we introduce a computer into the group is that a neutral act? There is a growing body of evidence that suggests not. It shows that children are more likely to work co-operatively and often achieve collaboration, when working on computer tasks rather than standard classroom tasks (Clements & Nastasi, 1985; Hawkins, Sheingold, Gearhart & Berger, 1982; Underwood & Underwood, 1990).

A second important controlling variable of group effectiveness is the gender composition of the group. Our own studies have shown girls collaborating more readily than boys and showing commensurate learning gains. These observations from classroom activities are reflected in the results from our controlled classroom experiments on the effectiveness of single gender and mixed gender pairs of children. In a series of studies we have used a Cloze task delivered with the INFANT TRAY program, in which a short passage of text has a number of letters replaced by hyphens. The task is to complete the passage by moving the cursor and entering possible letters with the keyboard. Successful attempts result in the letter replacing the hyphen. An example of such a passage, as it first appears in one of our studies, is shown at Figure 1. The completed text is:

> There once was a boy called Ethelred who was ready for anything. He lived in a town right next to one of those big jungles full of crocodiles, tigers, ruined cities and stone statues covered with vines.

Figure 1

THERE WAS ONCE A B-- C--L-D
ETHELRED W-- W-- RE--Y F--
A--T----. H- L-V-D I- A T---
RI--- N-X- T- O-- O- T--S- B--
JU----S F--L O- CR--OD---S,
T-GE--, RUINED C----S A-- S--N-
STATUES C-V--ED W--H V-N--.

The mixed pairs in our first study (Underwood, McCaffrey and Underwood, 1990) performed less well than the single gender pairs, using task performance measures such as the number of letters and words attempted in set time interval, and the number of letters and words that match those from the original passage. It seemed to us, in informal observations, that the mixed pairings were sharing the task in a different way to the all boy and all girl pairs. The mixed gender pairs were alternating their use of the keyboard, the single sex pairings were spending more time discussing the problem and thinking about the same part of the problem together. That is the mixed pairs were co-operating at an organisational level, turn-taking, but were not engaged collaboratively on the project.

A similar language-based study reported by Fitzpatrick and Harding (1993) adds to this description of mixed gender pairs. Using a word completion task they again found that all boy and all girl pairings co-operated together negotiating joint actions. The mixed pairs worked asymmetrically, with one partner dominating, and usually the boy taking control of the keyboard. The single gender pairs also worked faster at the task. Howe, Tolmie, Anderson & Mackenzie (1992) have also reported interaction failures for mixed gender pairs in comparison with single gender pairs working with physics problems. Their mixed pairings talked less, and problem solving success was related to specific characteristics of dialogue. The successful pairs tended to talk about possible explanatory factors and predictions in a problem concerning the paths taken by various falling objects.

As we have already been shown, working in a group is not necessarily easy for children who have been given little guidance or encouragement to do so (Brown, 1988; Cowie *et al.*, 1994). In a second TRAY task with similar groupings of children, half our sample children were explicitly instructed to work co-operatively with a partner and half were told that their individual contribution was most valued (Underwood, Jindal & Underwood, 1993). The mixed pairs continued to perform less well than single gender pairs, and they also showed little benefit from the instructions to co-operate. All girl pairs co-operated with whatever instructions they were given. Interestingly, the largest performance gains were shown by boys in single sex pairings who had been told to work together. These boys had changed from a previously competitive strategy to a collaborative one with commensurate gains. The benefits of the instructions to collaborate were seen in pairs who discussed the problem and agreed upon a joint action. This concurs with Johnson, Johnson & Stanne's (1985) finding which showed that there were significant educational gains in terms of attitude and performance in groups who were working together.

An analysis of the interactions between the children in our groups was conducted to establish the level of co-operation or collaboration between the children. Each statement of the dialogue between children as they negotiate their way through the task has been categorised according to the Bales schedule that breaks discussion into four main types: group agreement, offers suggestions/answers, asks for suggestions/answers, and group disagreement.

We have found a number of differences between the pairs using this analysis (Underwood and Underwood, 1995) and this is pointing towards an explanation of some of the performance differences. For example, and in the version of the TRAY task where all pairs of children were told to collaborate, the all girl pairings offered more suggestions than the other pairs, particularly offering more evaluation and analysis. Mixed pairings offered negative socio-emotional comments far more often than the other pairs, suggesting that they were disagreeing with each other more, and showing tension and antagonism. These results are in general agreement with other observations in the literature, with mixed groupings tending to talk less about the task and with one partner dominant, especially with the boy partner taking control of the keyboard/mouse.

What happens if we change the task?

One key issue concerning these gender findings is whether they are task specific. Infant TRAY is a language based program that might be assumed to favour girls over boys. Fitzpatrick and Hardman (1993) were also working in the language domain. If our findings are restricted to such a task then, although they remain interesting, they are of far less significance.

The question of the robustness of our findings was raised by an early LOGO study. The Hughes and Greenhough (1989) study of paired working with a programming task and a floor turtle produced contrasting results to our own. Their programming task found that all girl pairs performed poorly in comparison with both the boy only and mixed pairings. Hughes and Greenhough summarised their results as showing that whenever there was a boy present in a pair, then programming performance was good, but all girl groups were at a disadvantage. We questioned whether this was a task effect; LOGO is a spatial activity that we might assume boys would warm to especially as the activity also used a robotic toy. Equally this study was conducted with children three years younger than in our own studies.

Yelland's (1995) attempted replication of the Hughes and Greenhough result is interesting. Although she was initially successful in replicating the findings, she went on to show that the poor performances of girl only pairings disappeared over time and equally significantly depended upon the criteria for success. Girls were slower but less error prone, and on this basis could be considered to be more successful than the boy only pairs. As Schwartz (1995) has emphasised the nature of the outcome measures can have a significant impact on the results of such studies.

Lee (1993) working with children of 9 to 12 years of age, in groups of four, using the adventure game WHERE IN THE WORLD IS CARMEN SANDIEGO? also found major differences in the nature of the interaction between the children in single sex groups. In girl only groups there was increased overall interaction, a willingness to ask for help and just as importantly a willingness to provide that help. This was not so for the all boy groups, but the boys in mixed gender groups talked more and asked for and received support from the group. Disturbingly the girls in mixed gender groups talked less and, in male dominated groups, were unlikely to receive help from the group. Furthermore, any help that the girls in these groups did receive tended to be inadequate.

A final important point is that girls made far fewer negative emotional statements than boys. Boys were particularly prone to making negative comments in all boy groups; when a girl was present the number of negative comments declined and in the female dominated group no negative comments were made at all by either sex! This contradicts one of our findings but can be explained by the size of the groups. In mixed pairs the boy can dominate but it is less easy for a single boy to dominate a mixed foursome. The girls gain strength from each other.

The observation of children's discussions while they were programming with KidSim, a graphical programming environment, showed that task-oriented discussion predicted performance (Underwood, Pheasey, Underwood, and Gilmore, 1996). The total amount of time necessary to complete a given programming exercise showed a simple correlation with the number of statements per second in which the pairings gave opinions or analysis, expressed feelings and wishes, but the regression analysis also highlighted the importance of tension-releasing statements on good performance. As for the TRAY task, statements of antagonism lead to reduced performance in the form of increased time to solve the problem at hand. However, the most successful account of performance was with rule testing in which pairs who felt able to express opinions, to analyse the situation and express agreement and understanding outperformed other groups. The clear result from this study is that when children work in groups around a computer their discussions are task-oriented. The amount of time spent testing programmed rules was predicted almost entirely by two categories of dialogue, the number of opinions expressed and the number of affirmative comments. Very little of the discussion during rule testing concerns anything but the evaluation of those rules. Offering opinions and suggestions also accounted for most of the variance in total programming time, and in time spent writing the rules. The common feature of these relationships is that the discussion is reflecting the needs of each task.

The findings in the KidSim study are very similar to those of our TRAY studies and of Lee's adventure game. In these three problem solving environments, pairs who talked constructively together, introducing knowledge and ideas to the other member and accepting information from their partner, out-performed those who did not. In the TRAY environment this was more likely to be girl only groups, although boys would and could co-operate if encouraged to do so. Mixed gender pairings were not able to interact. Organisational constraints meant that there were no mixed gender pairings available to the KidSim study.

The tasks discussed so far are all problem solving but what happens if the task itself does not require a collaborative mode of working? At first sight working in groups appears less likely to affect children's learning with less problem-oriented but equally interactive programs such as LIVING BOOKS (a multimedia reading development application).

We set children to work in pairs using the Bederbund CD-Rom storybook 'Arthur's Teacher Trouble' by Mark Brown (Underwood and Underwood, 1996). The screen-pages of the storybook follow the main character, Arthur, through a school spelling competition. At the start of each screen-page part of the text of the story is displayed and read aloud. In addition, a rich illustration of an appropriate part of the text is displayed which the reader can interact with by pointing and clicking on different aspects of the display. Clicking on a word results in its pronunciation, either in isolation or together with other words. Clicking on a feature of the illustration results in an animation.

The 'Arthur' program is highly interactive but less problem-oriented than our other tasks. An analysis of the dialogue between the pairs of children revealed that all girl pairings showed more tension release by joking and laughing than the other pairs. Pairs of girls also asked each other for information. This more collaborative interaction by the girls appeared not to have any performance advantages, however; a result at variance to our findings in numerous TRAY studies and in the KidSim study. One explanation could be that the task itself does not benefit from group collaboration, and that the main interaction is between child and screen rather than between members of the group. Further analysis of a story recall task that took place three weeks after completion of the intervention, did reveal surprising performance gains by the girls, however. The recall of story events during free recall was especially noticeable in the stories written by girls who had been in all girl pairs (see Story 1). Furthermore, when girls worked on the storybook with boys their free recall was on a par with that of the boys; that is, it was depressed compared to girls in the all-girl pairings. Stories by boys from mixed and single gender pairings tended to show little content knowledge (Story 2). Accounting for this effect is not straightforward. The stories were written by individual children working in a quiet classroom. The CD-Rom storybook interactions were completed several weeks previously, and yet some aspect of the paired interaction appears to be associated with recall of the storybook.

Yet, despite the nature of this task, our analysis did reveal differences between the gender pairs while they were working with the storybook and this may provide a clue as to why girls' memories are influenced by their working partner. Pairs of girls joked and laughed more than other pairs, and were also more likely to ask each other for information than the other pairs. They were more relaxed about working with a partner, they regarded their partner as a source of advice and information, and this had an outcome in their superior recall of the storybook. It would appear that even in tasks with limited cognitive load, working together can have beneficial effects.

Two examples of the post task free re-call writing task

Story 1: A typical free recall story showing knowledge of the salient points of the original Arthur story

One morning <u>Arthur</u> woke up. 'First day at <u>school </u>today' he said. When he got to school he sat in a place on the third row just then <u>Mr</u> <u>Ratburn</u> came in, I don't whether you know but we have got a <u>spelling test</u> with <u>100 words</u> in, we will start today but over the term I ill want you to find other words at home. When Arthurs went home that night he started trying to <u>find more words</u>. The next day when it was time to go home Mr Ratburn told everyone to line up in alphabetical order and <u>go out one by one</u>. after <u>all the other clss's had gone</u>. Time went by and Arthur had collected lots of words. Then the day of the <u>Spellathon</u> came, everyone was sitting in there placces <u>waiting for the reults,</u> '<u>Arthur</u> & Robert are the winners, come forwad'. said Mr Ratburn. Arthur blushed. when he went home he told <u>mum & dad</u> & said to <u>his sister,</u>' our luck you don't have to do any spellathon do DW'.

Story 2: A typ ical free recall story showing use of key characters but a new story-line.

One day there was a Boy called Arther and he had a sister who is alway's a pain in the neck and his sisters name is D.W. One day arther was sitting in his sitting room and he saw some-thing strange he had a look to see what it was and he coulden't Beleeve his eyes he creeped out of the door and saw a strange animal with 4 eyes to things like horns and the was to less 4 arm's and he went out of the door and said hello and he said the same and he said would you like a ride arther said yes and he closed the door and weant out sid he said to the Planet smiky

Underlined phrases show awareness of the story line.

In conclusion

Children working in groups have significantly different experiences depending on their gender and the group composition. There may also be an influence of task. The recurring observation from natural classroom tasks is that boys see the computer as being in their domain, but classroom experiments find that in single gender groups the girls perform as well if not better than boys. In our experience it is only when boys and girls are paired together that they perform poorly. Kruger's analysis of collaborative styles helps us to understand the dynamics of group interactions and their effects upon thinking skills. Her studies of paired reasoning showed that the nature of the discussion influenced subsequent performance. Pairs that offered suggestions to a problem, and considered each other's suggestions, did better in an individual test of reasoning administered later. In particular, it was the consideration of rejected solutions that was associated with the development of thinking.

Discussion provides the opportunity to compare ideas, and to evaluate suggestions, something that our all girl pairs were doing with little encouragement from us. Lee's girl foursomes were equally effective. In the case of our mixed pairs there was little discussion and unimpressive performance. The most effective thinkers justify their own ideas and also take account of their partner's suggestions, with the eventual rejection of failed solutions being the best indicator of new understanding. Now that we know what current patterns of performance look like, and how children tend to discuss problems when working around a computer, we can encourage the kind of discussion that will result in the development of their powers of thinking. When we have observed mixed pairs working successfully together (Underwood & Underwood, 1995) each child has been comfortable in offering suggestions and in analysing and evaluating each other's suggestions.

Does the nature of the task matter? Those researchers arguing the case for group work have done so on the grounds of encouraging some form of cognitive conflict and as a support for problem solving activities. We have found from our exploration of 'Living Books' that working together may also benefit lower level information retention and recall, that is the benefits of group work are not confined to higher order problem solving activities. Equally, Dunne and Bennett 's assertion that group work should start from tasks with low cognitive load is not borne out by our findings. We have found simply telling children that the goal is to co-operate can affect boy only groups even when working on high level problem solving tasks such as our INFANT TRAY Cloze task, although this is not true for mixed gender groups.

References

Alexander, R. Willcocks, J. & Kinder, K. *Changing Primary Practice.* Falmer Press: London, 1989.

Bales, R. F. *Interaction Process Analysis.* Addison-Wesley: Cambridge, Mass, 1950 .

Biott, C. 'Co-operative Group Work: Pupils' and Teachers' Membership and Participation'. *Curriculum,* **8**: (1987), *5-13.*

Brown, A. & Palincsar, A. *Guided Co-operative Learning and Individual Knowledge Acquisition,* Technical Report 372. Bolt, Beranak and Newham Inc: Cambridge, Mass, 1986 .

Brown, R. *Group Processes: Dynamics within and between Groups.* Basil Blackwell: Oxford, 1988 .

Clements, D.H. & Nastasi, B.K. 'Effects of computer environments on social emotional development: Logo and computer aided instruction'. *Computers in the Schools,* **1**: (1985) 11-31.

Cowie, H., Smith, P., Boulton, M. & Laver, R. *Co-operation in the Multi-ethnic Classroom.* David Fulton Publishers: London, 1994 .

Dunne, E. & Bennett, N. *Talking and Learning in Groups.* Macmillan: London, 1990 .

Fitzpatrick, H. & Harding, M. 'Girls, boys and the classroom computer: an equal partnership?' Paper presented at the meeting of *Developmental Psychology Section of British Psychological Society,* Birmingham: September 1993.

Galton, M., Fogelman, K., Hargreaves, L. & Cavendish, S. Final Report: *The Rural Schools Curriculum Enhancement National Evaluation (SCENE) Project.* Department of Education and Science: London, 1991 .

Hawkins, J., Sheingold, K., Gearhart, M. & Berger, C. 'Microcomputers in schools: impact on the social life of elementary classrooms'. *Journal ofApplied Developmental Psychology,* **3**: (1982) 361-373.

Hill, G.W. 'Group versus individual performance. Are N+1 heads better than one?' *Psychological Bulletin* **91**: (1982) 517-539.

Howe, C., Tolmie, A. Anderson, A. & Mackenzie, M. 'Conceptual knowledge in

Physics: The role of group interaction in computer-supported teaching'. *Learning and Instruction* **2**: (1992) 161-183.

Hughes, M. & Greenhough, P. 'Gender and social interaction in early LOGO use'. In Collins, J. H., Estes, N., Gattis W. D., & Walker D. (eds) *The Sixth International Conference on Technology and Education,* Vol.1. CEP: Edinburgh, 1989.

Johnson, R. T., Johnson, D. W. & Stanne, M. B. 'Effects of co-operative, competitive, and individualistic goal structures on computer-based instruction'. *Journal of Educational Psychology* **77**: (1985) 668-77.

Kruger, A. C. 'Peer collaboration: conflict, co-operation or both?' *Social Development,* **2**: (1993) 165-182.

Lee, M. 'Gender, group composition and peer interaction in computer-based co-operative learning'. *Journal of Educational Research* **9**: (1993) 549-577.

Light, P. & Colbourn, C. 'The role of social processes in children's microcomputer use'. In Kent W.A. & Lewis R. (eds) *Computer Assisted Learning in the Social Science and Humanities.* Basil Blackwell: Oxford, 1987.

Piaget, J. *The Origins of Intelligence in Young Children.* IUP: New York, 1952.

Reid, J., Forrestal, P. & Cook, J. *Small Group Work in the Classroom, Language and Learning Project.* Education Department of Western Australia: Perth, W.A., 1982.

Schawtz, D.L. 'The emergence of abstract representations in dyad problem solving'. *The Journal of the Learning Sciences* **4**: (1995) 321-354.

Schoenfeld, A.H. 'Ideas in the air: Speculations on small group learning, environmental and cultural influences on cognition and epistemology'. *International Journal of Educational Research* **31**: (1989) 71-88.

Slavin, R. *Co-operative Learning in OECD Countries: Research, Practice and Prevalence.* Centre for Educational Research and Innovation, Organisation for Economic Co-operation and Development: Paris, 1993.

Teasley & Roschelle 'Construction of a joint problem space.' In Lajoie, S.P. & Derry, S.J. (eds) *Computers as Cognitive Tools.* Lawrence Erlbaum Associates: Hillsdale, N.J., 1993.

Underwood, G. 'Collaboration and problem solving: Gender differences and the quality of discussion'. In Underwood, J. (ed) *Computer Based Learning: Potential into Practice.* David Fulton Publishers: London, 1994.

Underwood, G., Jindal, N. & Underwood, J. 'Gender differences and effects of co-operation in a computer-based language task'. *Educational Research* **36**: (1993) 63-74.

Underwood, G., McCaffrey, M. & Underwood, J. 'Gender differences and effects of co-operation in a computer-based language task'. *Educational Research* **32**: (1990) 44-49.

Underwood, G., Pheasey, K. Underwood, J. & Gilmore, D. 'Collaboration and discourse while programming the KidSim microworld simulation'. *Computers and Education* **26**: (1996) 143-151.

Underwood, G. & Underwood, J. 'Gender differences in children's learning from interactive books'. In Robin, B., Price, J.D., Willis J. & Willis, D.A. (eds) *Technology and Teacher Education Annual* AACE: Charlottesville, Va., 1996.

Underwood, J. & Underwood, G. *Computers and Learning: Helping Children Acquire Thinking Skills.* Basil Blackwell: Oxford, 1990.

Underwood, J. & Underwood, G. 'When do groups work? Effective collaborations with classroom computers'. *NFER Topic* **14**: (1995) 1-6.

Vygotsky, L. *'Mind In Society: The Development of Higher Psychological Processes'.* Harvard University Press: Cambridge MA., 1978.

Yelland, N. 'LOGO experiences with young children: Describing performance, problem solving and social contexts of learning'. *Early Childhood Development and Care* **109**: (1995) 61-74.

Picture Information Literacy

Stephen Marcus

> The...present work will be executed with the greatest care...and the scenes represented will contain nothing but the genuine touches of Nature's pencil.

Photography provides a particularly rich arena for 'learning where the truth lies', which as designer Robert Mohl has noted, includes both learning where the truth resides and learning where it deceives.

The quotation above is taken from an advertisement for *The Pencil of Nature*, the first published photographic book. There is a tradition, which this book helped establish, of equating photography with truth ('the camera doesn't lie'), but in point of fact, photography 'despite its apparent simplicity, constitutes a rich and variegated language, capable, like other languages, of subtlety, ambiguity, revelation, and distortion' (Richtin, 1990).

As a technology, however, photography or the camera is no longer a simple matter (not that it ever was to a large percentage of the population, hence 'point and shoot' and 'disposable' cameras). Alternative film formats and digitalisation have been accompanied by attendant developments for making, altering, storing, retrieving, and disseminating information as images, including computer disks, CD-Rom, laserdiscs, and Internet/World Wide Web environments. The camera as a tool and medium has merged with the computer, creating an even more compelling, and disconcertingly complex set of challenges and opportunities for discovering and exploring the traditional distinctions between data and information, information and knowledge, knowledge and wisdom.

In educational settings, where such distinctions form the basis of curriculum design and pedagogy, the renewed interest in the concept of multiple intelligences (Gardner, 1983) has been accompanied and informed by the growing application of more varied strategies for developing a wider range of student (and teacher) aptitudes and abilities. In addition, the extensive array of new technologies provides both students and educators with new powers and incentives for showing the world what is on their minds.

The emphasis here on images is also grounded in certain 'assumptions about the development of literacy and the making of meaning.... [That] we use images, often mediated by language, to make sense of our world, and this activity resides at the core of thinking.... [That] images are natural parts of any discourse community and [are] thereby highly intertextual.... [And that] images are rhetorical [and] shape how we think about the world' (Fox, 1994). In other words, as students 'begin to see better, they begin to think and write better' (Hovanec and Freund, in Fox 1984).

As usual, however, merely getting a tool and learning how to operate it does not ensure that it will be put to its most effective use. The discussion below indicates how even an old and relatively simple photographic technology can help us increase our 'visual IQ' and how writing can help give a voice to this intelligence so that students and teachers can both understand and use the 'rich and variegated language' that Ritchin ascribes to photographs, one that allows us to explore our world 'as much through the ambiguity of metaphor as through the sharpness of the lens, the display of visual fact'.

The discussion to follow includes a variety of approaches for combining photography and writing in an effort to increase students' and teachers' visual information literacy. The first section describes a project in which a small group of teachers conducted classroom-based action research in an effort to document and understand their own lives in their classrooms. The second section describes a much larger project, in which some 6000 U.S. teachers at 46 sites of the National Writing Project participated in workshops focusing on visual literacy and learning.

1. Visual information re: search for self

Send us a picture worth a thousand words – along with the thousand words.

This was the final assignment that was given to a group of 35 teachers from around the United States who participated in the Teacher-Researcher Instant Photography (TRIP) Project, a special undertaking growing out of a strategic alliance between the National Writing Project and the Polaroid Education Program. The NWP/PEP Alliance is an on-going effort to combine writing with the development of visual literacy (one of the 'multiple intelligences') that is to use photography to help people show what was on their minds and to add eyes to their voices.

The TRIP project was designed to affect and enrich the participants' understanding of themselves as professionals and their understanding of their students. It was grounded in the notion of the 'reflective practitioner', an approach that emphasizes self-analysis and self-assessment and suggests that professional development programs need to take teachers' own beliefs into consideration, including their implicit assumptions concerning what learning is, how classrooms function, who their students are, and the subject matter that is to be taught (Schön, 1987). The TRIP project sought to help teachers combine writing with a visual medium in order to compose and communicate who they are, whom they serve, and what they would like themselves and their students to become.

The TRIP project was established as a special-interest group within the larger NWP/PEP Alliance. The project helped teachers combine visual literacy and writing as they developed professional portfolios that included 30 pictures that answered the same basic question: 'Who are you as a teacher?' The participants got a special TRIP Kit that included film, background reading, a special journal (with a glue-stick for attaching photos), and a series of structured but open-ended assignments. The teachers used cameras they had already received after attending NWP/PEP workshops. Their responses to the assignment noted above ranged from research reports, to HyperCard stacks, to deeply introspective narratives, to rhyming couplets.

The project's approach emphasized the distinction between 'taking' and 'making' pictures. The former is a more passive activity. The latter requires conscious and informed choices in the construction of images. A related distinction is between pictures *of* people and pictures *about* people (or things). In this regard, there is a helpful rubric for identifying and assessing three levels of photos. A Level One photo is one in which the thing you wanted to 'take a picture of' actually turns up within the borders of the photo. It's 'in the frame'. A Level Two photo is one in which the thing you wanted to take a picture of is in the frame, and it's appropriately featured, centered, and in focus. A Level Three photo satisfies the demands of Levels One and Two. In addition, it reveals an interesting or unusual perspective on its subject matter. The picture is not just *of* something, it's *about* something.

It's possible to adapt these criteria to the assessment of writing samples: a Level One sample actually attempts to complete the assignment; a Level Two piece, in addition, focuses appropriately on the assignment and demonstrates suitable degrees of fluency, form, and correctness; a Level Three piece meets the previously described criteria and also takes a fresh or noteworthy perspective on the assignment or topic. Explicitly utilizing analogous rubrics in two ostensibly different media is one simple way to begin integrating photography with writing. It's also possible to compare prewriting, writing, and rewriting with previewing, viewing, and reviewing, thereby exploring the process of making a visual product (an image) rather than a written one. These and other perspectives were part of the TRIP project.

Participants were provided with a variety of strategies for previewing, viewing, and reviewing their investigations. Here are just a few of the suggested assignments. Teachers were encouraged to interpret them in whatever ways would make the time spent valuable to themselves. (Note that the assignments could easily be adapted to student-oriented projects aimed at answering the question, 'Who am I as a student?'):

- make a set of pictures that tells a story of who you are as a teacher. Make notes that go with the pictures
- take a picture of who you are as a teacher that's a first impression. While it's developing, write about what you *think* it will show. After it's developed, examine it and write about what it *does* reveal. Wait one week. Write about it again

- take three pictures that answer the question, 'Who am I as a teacher?' Put them in order of importance. Explain yourself in writing
- pick a student to focus on. Take three pictures that answer the question, 'Who is this student relative to me?' Put them in order of importance. Explain yourself in writing
- take three pictures that answer the question, 'What do I really teach?' Put them in order of importance. Explain yourself in writing
- have three of your students take one picture each, pictures that show you as a teacher from the students' perspectives. Before you discuss the pictures with the students, write your own explanations of each picture. Then talk with the students. Then write some more
- answer the question, 'Who are you as a teacher?' with three pictures that do not have any people in them. Examine each picture and write down five words that help explain what the picture reveals.

Photo-prompts such as these suggested but did not require certain avenues for self-reflection. Since the participants had to provide only one photo as a 'deliverable' (along with the 1000 words), they did not have to account for or document their various 'first draft' photos. Suffice it to say that no one reported having left-over film.

There is often, of course, a reciprocal relationship between teachers' studying of their teaching and the evolving conduct of that teaching. Instruction provides a test-bed for investigation, which then informs changes implemented on the basis of that research. Such was the case, for example, in work one of the TRIP participants was doing with aphasic patients. Ellen Bernstein-Ellis helps treat individuals with mild to severely impaired language processing abilities. In so doing, she is looking for 'any spark of communicative motivation that feels genuine. Genuine in the sense that there is a real reason to convey or respond to something'. (Quotations are taken from her 1000-word contribution 'Snap, Flash, Spark: Some Reflections on the Evolution of the Photo Event Journal'.) Bernstein-Ellis was practiced in the traditional forms of didactic therapy: 'In treatment exercises, I ask questions to which the patient knows I already know the answers. It may be good to practice this type of communicative task, but...it is not a substitute for communication'.

Bernstein-Ellis began to let patients and families document holidays and other events and to create photo-stories about those events, with a beginning, middle, and end. The inclusion of these photos in event journals 'eased the discomfort that family and friends experience when trying to carry a large share of the communication burden....By letting the story unfold through the...pictures, the conversational format changes from primarily a guessing game to exchanging social comments...in context [which] are frequently processed well by aphasic individuals....Either way there has been a two-way exchange instead of a one-way drill for information'.

She also notes that an additional 'lightbulb flashed in my brain...in addition to that of the cameras'. She realized that her patients 'were spending almost half of their day twice a week at the Easter Seals Center, yet they had very little ability to convey to their family or friends what had happened in the course of any session'. Using photo journals, including pictures and key words, the patients were able to take home a record of events. A small pilot study suggested that 'the journal helped to improve the qualitative and quantitative output of the aphasic individual, and it shifted the format of the spouse's conversation from one of primarily questions to one of comments'.

This teacher's involvement with the TRIP project had important pay-offs for her patients and for herself as a therapist/teacher. She notes that something 'that promotes a genuine communicative spark has great clinical value. I have found that the click and flash from the...camera is followed by a communicative spark. I hope to learn more about that spark'.

In contrast to the relative paucity and difficulty of expression inherent in the previous example, Ruth Ann Ritter's 1000-word report begins with a display of 'onomatopoeic pandemonium' ('EWWW, UGgH & OHhhhh') that is familiar to many teachers who try to interest their high school students in poetry. To help address this problem, Ritter adapted a strategy from the movie *Dead Poets' Society*, allowing students to stand on (and jump off) a desk in the course of reciting and talking about poetry (Figure 1).

Figure 1. 'Dead Poets' Society'

Ritter used her camera to document these displays, and significantly, such an event became a defining and transforming moment for her. 'Taking the picture of Kevin's presentation was a powerful moment for me, too.... It was as if the other thirty-seven students did not exist at that moment. Kevin was the only thing in my frame of reference; I was not distracted by Brynner's incessant hand-tapping; I wasn't busy reprimanding

Josh for chewing gum in class, again. All that mattered was in the view of the photographic eye. This reminded me of my need to focus on my students with more *intent*'. Ritter adds that she is too often focused on the other students and 'what they are doing or not doing rather than the student who is in front of the class with something valuable to contribute.....I need to find a balance between focusing on one student and all students. Much like the students who stood on my desk, I, too, must constantly look at my classes from different vantage points, through different lenses'.

An anthology of teachers' 1000 Words (along with their selected pictures) has been distributed to TRIP project participants (two of the selections are included below). The thoughtfulness evidenced in those documents provides a useful unobtrusive measure of the value of a camera's viewfinder as a tool to help teachers frame and focus their attention, both with an outward gaze and inward reflection. As one teacher put it, 'I notice a shift in my thinking about "Who I am as a teacher" from literal to more metaphorical—[fewer] pictures [of the] classroom, more symbolic ones now'.

The TRIP project's mission was to support a group of educators who were willing to explore, document, and communicate insights into their professional lives. The project was embedded in a context that emphasized the importance of people's constructing information and meaning out of the raw material of image and language, of transforming data into information, information into knowledge, and knowledge into wisdom. In addition, the project was based on an increasingly common redefinition of the teacher's craft and role. Teachers who observe, question, and analyze their own classrooms find that such classroom-based research benefits both themselves and their students.

Our stories, our selves[1]

Alice Slusher

Is this a story or a question? Remember, we just don't have time for your stories.

A museum guest-speaker is explaining to my fifth graders her ground rules for keeping the discussion focused. She was on a tight schedule and had so much material (on endangered species, ironically) to share with our class.

Since then, listening to similar pronouncements from others, I find I have many questions. When can our children tell their stories? How else can they show they're finding connections between their own lives and the lessons of the school day? How can I possibly hope to reach them without knowing who they are? How many of their stories can I bear to know this year?

And now, in seeking some response to the question 'Who am I as a teacher?' I sense that my Appalachian roots determine this answer more than I ever realized. Values many generations deep concerning community and place keep shading my perception of myself. My children's stories, my colleagues' stories, my school's stories keep crowding into my own. I dream these people and this place. I am not the same teacher this year that I was last year. Stress and distress in my school community are making me doubt answers of past years.

This school year has brought many days darker than this photographic image. In turning through my journal and looking for a defining shot, I noticed how many of my images are seeking sunshine. Students posing with just forced narcissus blooms, translucent silhouettes decorating the wet window sills, zigzagging shadows from the convoluted handicapped ramp out the upper level exit, close-ups of children engrossed with the stories they're sharing. Yet I keep returning to this one, shot just after the buses pulled out of the parking lot on a brilliantly sunny November afternoon. The up-ended chairs are a simple gesture to make our custodian's after-school sweeping job a little easier. The reflection off their legs could also be a signal of distress. This was Thanksgiving week, a week of family and blessings.

Jay's desk is the second one in the photo. It has been unoccupied all week. This darkness is preserving the day of his father's funeral. We are engulfed with the pieces of Jay's story. That he loved spending Saturdays working with his daddy. That his father spent a week and a half in neurointensive care. That Jay's family has no medical insurance. That Jay couldn't stand to come to the funeral home the previous night. What we don't know is Jay. How will he be, who will he be, when he comes back to school next week. We make cards and talk about what we could say or give to Jay that might help. 'Your heart talking to Jay's heart', I offer. Later, as I read each note, the darkness abates for awhile. Each child has understood, truly understood.

This is not a year for universal understanding. My 21 students include 11 who qualify for special programs and others for whom labels have just not stuck. The stair-stepping heights of the desks in this photo hint at the diversity of levels and talents present. In the fragmented schedule that comes with special needs, there is a single half-hour block once a week when all the students are with me. I have students who know shelters and courtrooms, students who know sexual abuse and emotional desperation. I can't begin to imagine all their stories. I am overwhelmed by all the questions. I have learned not to assume that even the most basic concepts or directions are clear. Wanting the usual flares of excitement and joy most 10-year-olds bring to their learning, I struggle against my own feelings of failure. Poems that sparked whole anthologies last year, questions that germinated into whole weeks of study drop dull and timeless. I look into this darkness in somewhat the same way I look into the depths of the hemlock outside my bedroom window, checking to see if there is silent rain falling. I search for the light against the dark. Are there writers growing and quiet voices gaining confidence? Some days I strain and watch, and seeing nothing, I must simply hope.

The shortest desk of this grouping is the one I spend the greatest amount of time visiting. It is homebase for Mitch. So much energy and so much rage and so much charm in so tiny a body. Some days, he's all over the room, accusing a friend of stealing his watch, begging a girl across the way to 'go with him' and dancing as she agrees, waving his hand and yelling for my immediate attention only to refuse eye contact or assistance when I move to him. Some days, he's so filled with questions about what's going on with his family that he can't even stay in our classroom, racing down the hall, always looking back to see if anyone cares enough to come after him. Some days, he's sure his spelling grade is determined solely by the color of his skin. Some days, he's a little boy who won't let go of a hug.

Looking again at the photograph, there are metallic flashes reflecting and desk tops edged with lights. Still, it is the darkness of floor, ceiling, beyond the chair leg spaces, that dominates. My school is in the process of being remolded. Single light bulbs dangle amidst the exposed pipes and wires that form our hallway canopies. Our students wind their ways through hallways half blocked by ladders and spools of electrical cable. Drills and hammers punctuate our lessons, layers of dust blanket our shelves. We must endure noise and cold and dirt and closed restrooms and flooded closets and chaos at least through the end of this school year.

Our faculty is in a sort of remolding stage, too. Teachers used to spend hours after school working together on projects, talking about the day's high and low moments, offering suggestions and support. Candy bars and cartoons in mailboxes, classes covered during emergencies, compliments and cheers for one another's successes were commonplace. Now, I listen to a 25-year-old professional confess that for the first time in her career, she dreads even walking into our building. Another strong teacher who's worked closely with me for 8 years gets damp eyed as she tells me she never knew how hard it could be to make herself not care, not do more than the minimal requirements. I jam another book into my stuffed bag and hurry out the door. How can a community become so damaged in such a short time? We have new administrators, with new agendas. Changes have been dictated by people who don't know our stories, who we are, what we do, what we need. We have been blinded by little deceptions and faulty communications. The pain of individual betrayals spreads to our greater community leaving us all grim and wary.

Tessa is my newest student. She came to us from an urban school with a confidential file an inch thick. Her stories were of severe behaviour problems, severe learning problems, severe family problems. Her first words to me were, 'I didn't learn a thing last year. I don't know why they passed me'. This is not the Tessa sitting in the fourth desk now. This year's story has been of a child who loves music and art, who leaves me little notes and poems. She's found friends here. I smile at her eagerness and tuck her notes away, forgetting those first questions I had.

Beyond the desks, the photograph reveals shelves of books barely noticeable if not for the gilt lettering on the encyclopaedia set. Our big fifth grade trip is always to Monticello, Thomas Jefferson's home. Every year's tour leaves us fascinated and awed. Once, the chosen day was drizzly and overcast. Passing through rooms illuminated only by skylights and windows, the children were a little disappointed by the dimness. As we moved into Jefferson's library, the guide pointed out the gold leaf detailing on the books and explained its value in locating particular volumes on just such mornings centuries past. The children flashed glances in my direction to see if I shared their delight. The delight of knowing such lights sustains me.

Who am I as a teacher?

Jinny Woodall-Gainey

Bridge (brij) n. 1. A structure spanning and providing passage over a waterway or other obstacle. (American Heritage).

Who am I as a teacher? I am a bridge, providing passage for my students from the knowledge they now possess to what they have yet to learn.

'Learner (lurn-er) n. 1. One who gains knowledge or understanding of or skill in by study, instruction, or experience' (Webster's)

Who am I as a teacher? I am a learner, striving for focus, trying to sharpen the image in the camera as I hone in on the answer to the question.

I am a bridge....

I provide passage for Meghan from the first draft of her poem, which is filled with vague lines about children and how sweet and fun they are. I don't give her the specifics, simply a way over the obstacles of her fuzzy thinking. I ask her instead to see the children in her mother's kindergarten class, to show them to us, and she does. Her final draft has them making a messy snack, telling secrets, and finger painting. Her words sing.

I am a learner....

I ponder who I am as a teacher, and I wonder what image in a photograph will show that. Words are much easier for me than visual images. I write, I think, I read, and finally—timidly—I am the camera and view the product. I am surprised at the shadows that obscure my students' faces, and I try again from a different angle where the light is much better. Faces are lit up now but off center. There is too much to remember for this novice!

I am a bridge....

With Virginia Tech professor Kathleen Carico, I am a bridge between communities of learners: my middle school students and undergraduates in Kathleen's Adolescent Literature class. Once a week I see the excitement on my middle schoolers' faces as they call these future-teachers friends. I watch as the semester unfolds, as each becomes touched by the others' responses to The Chocolate War, Julie of the Wolves, Roll of Thunder, Hear My

Cry. I watch Andrew, who suffers from clinical depression, come out of himself each week in his letter to his Tech pen pal, Darrell. I watch the Tech students gain insight into the minds, the thinking, and the reading habits of middle school students. I am glad to be a bridge.

I am a learner....

I learn that observing kids from behind the camera lens gives me ideas about teaching in much the same way that listening to them and reading their writing gives me ideas. In the photo I see Rachel's interest in what Beth is reading to her, and am reminded of Beth's struggles to make her writing clearer. I wonder if showing her this photo, with Rachel so obviously interested in what she is reading, would help her in much the same way that the feedback Rachel gave her minutes later did. Observing my students, their needs come into clearer focus, and I learn where we need to go next.

I am a bridge....

I try to help students see that '.....writing is largely a process of choosing among alternatives from the images and thought of the endless flow, and this choosing is a matter of making up one's mind, and this making up one's mind becomes in effect the making up of one's self' (James E. Miller, 1972).

I am reminded of Houston, so excited at first about his story of a WWII fighter pilot, now struggling with how to bring it to a climax and a conclusion—and losing hope. He asks me what he should do. I fight the temptation to be more than a bridge, to give Houston specific suggestions and take away his right to discover for himself. I ask him to read what he's got, then I ask him to imagine how such a character would resolve the conflict, what could he 'see' him doing? What event might bring about a day of reckoning? Two days later, he brings me the polished product, much better than anything I could have offered.

I am a learner....

I read the letters, take more photos, re-read the [TRIP project] directions, take a few photos, think about the rubric for assessing the photos, try to remember "NO'O NO'O" [a TRIP project suggestion, from the Hawaiian term for a reflective state of mind] and try to choose from the images of the 'endless flow.' I begin to 'see' in a different way, much more aware of light and shadow, wanting to take pictures in places outside my classroom. Around the same time, I begin to realize that the answer to the question 'Who am I as a teacher?' may be more easily answered metaphorically than literally.

I begin to see possibilities: I spend an early morning trying to photograph the candle in the room where I sit quietly every morning in the dark, trying to simply breathe, simply be. I think of teaching as lighting the darkness. A nice idea, but the flash makes the effect impossible to achieve. But I have learned. And I know, after all, that the flow is endless.

I am a bridge and a learner...

So, after all, there are two images, one visual, one a concept, both captured in the photograph. The image of the bridge comes to me in class as I struggle to reach

Noah, a particularly troubled child with equal amounts of hostility and talent. Later I saw myself with Noah, trying to understand, knowing that without that connection there is no hope of making progress, and the image of the Walnut Avenue Bridge came to be. It was the metaphor I'd been searching for.

But could I photograph it? After several attempts, each one a little better than the one before, but none quite right, I thought of the other answer I'd been carrying in my head these many months, really my first response to the question 'I am a learner'. As a teacher, I am always learning: from my colleagues, from students, from projects like this one, from life. The photograph, I realize, shows both the bridge, and the fact that I have learned and am still in the process of learning.

And that's who I am as a teacher!

2. Picture this: some close-ups

By the end of its fourth year, the NWP/PEP Alliance will have provided one or more workshops integrating writing with visual learning for some 6000 teachers at 46 of the 160 sites of the National Writing Project. The various sessions have focused on incorporating visual literacy and learning with writing across the curriculum, student portfolios, and the teaching of writing, all the while enhancing students' motivation, creativity, and self-image.

In the current form, participants in 'supported' workshops receive approximately $75.00 worth of free materials, including cameras, camera bags, film, and curriculum materials. (Participants at NWP sites that have already sponsored one 'supported' workshop get everything but the film.) Along with the equipment and materials, participants receive a 3-hour in-service workshop that (as mentioned above) prepares them to help their students distinguish between pictures *of* things and pictures *about* things, between *taking* pictures and *making* pictures. Participants learn how to create and read pictures with increased sophistication and to understand how photography can help reach and enrich traditional curriculum goals. Teachers also learn how to use photography to help students tell their 'stories' in ways that reveal who they are and how they've grown.

Among other things, the NWP/PEP Alliance illustrates how tremendous value can be added to the classroom by putting even a 'low tech' tool into the hands of talented educators. What follows is a collection of events and applications that evolved out of the NWP/PEP Alliance using this old technology, along with some new sources and resources.

Our own NWP site is the South Coast Writing Project at the University of California, Santa Barbara. Its first visual learning workshop was led by Peggy

Hooberman, a former director of a non-profit multicultural arts organization who also had 18 years of classroom teaching experience. She was a terrific presenter with loads of amusing and instructive anecdotes. She had experienced every sort of problem that was likely to occur when using cameras in the classroom. She showed us how to recycle almost every part of an empty film packet (e.g. using the batteries for science projects and the different parts of the film container for picture frames). She made very clear the advantages and disadvantages of having various quantities of cameras in the classroom.

Toward the end of the workshop, I asked the teachers to take just a few minutes to jot down a classroom activity that combined writing with the use of instant photography. (They were to assume that they had a reasonable number of cameras, film, and other materials.) These 'bright ideas' were collected, printed in an anthology, and later mailed to the participants. They ran the gamut from ideas to use with student-teachers ('photograph scenes from around the school and write about them from a student's point of view') to activities for kindergartens ('make a floor quilt of pictures of small things on the floor, like staples, bits of confetti, a paper clip, etc., and write down what children think the objects are').

Even though the collection derived from our own workshop was extremely sketchy and very much of a 'first draft', it was in keeping with the Writing Project spirit that the teachers' own expertise be shared, and that by creating their own materials, they become more involved in the substance of the workshop.

A table of contents

I have, personally, always depended on the kindness of fortunate happenstance. In one case, I took a camera to a potluck picnic at the end of the first week of one of our Writing Project's summer institutes. I figured I'd take some snapshots of people to put on the bulletin board (another small way to help establish a sense of community). As it happened, I took a picture of the food table, groaning under the weight of various delicacies and outright food-crimes. While waiting for the picture to develop, I started chatting with one of the institute participants, who I knew was a science teacher. When the picture had developed, I looked at it and said, 'Um...here's a picture that doesn't have any people in it. Is there any way to use it for some science activities that also include writing?'

Could she ever. I later typed her comments and displayed them along with the picture in the institute meeting room. Here's a slightly edited version of what the teacher, Lyla Allen, suggested:

Food for thought

- start by sorting the objects on the table, 'living' vs. 'non-living'; polymers vs. non-polymers

- describe the sources of the objects. Tell the 'backstory', that is, how the objects wound up on the table. Pay attention to the geography, life cycles, and technology that brought them to market
- write their autobiographies. Take the point of view of a single molecule
- attend to the development of the object's nature or 'personality' (e.g., its sweetness or its degree of ripeness)
- tell the story of the object's future. Where does it go from here?

This was an impromptu activity that resulted in a brainstormed list of possible thinking and writing assignments. The photograph helped provide something concrete that could be handed over to other teachers or to students who wanted to pursue the activity.

What's in a name?

Here's another application. I was working with a group of English student-teachers, introducing them to the use of computers for the teaching of writing. I knew from experience that people often develop 'relationships' with their computers, often deriving from their feelings about what I've come to call 'the host in the machine'. I decided to make the implicit explicit. Before the students arrived for the workshop, I used a book of large tear-out masks and affixed a different one to each computer (*Fun Faces: 15 Punch-Out Masks,* by Pierre-Marie Valat. New York: E.P. Dutton). Thus, each computer was given a different personality. After the student-teachers arrived, they were asked to give their computers names and to write poems that elaborated and explored those names. They were then asked to take a picture of their masked computers.

Even though the students had only about 10 minutes to do their writing, they came up with some extraordinary personalities, ranging from Queen Rom-Rom ('With magic she draws us') to Sleeps Standing Up ('There was a time when he slept at night'), to Get Even (Figure 2), which was based on a computer with a pirate's mask.

Get Even

I've escaped, you see, from Disney's cage,

that computerized cacophony of fireflies and New Orleans

and boats with incessantly curious fun-seekers and new sweatshirts.

I am not angry with them,

but my name says what I have to say to them.

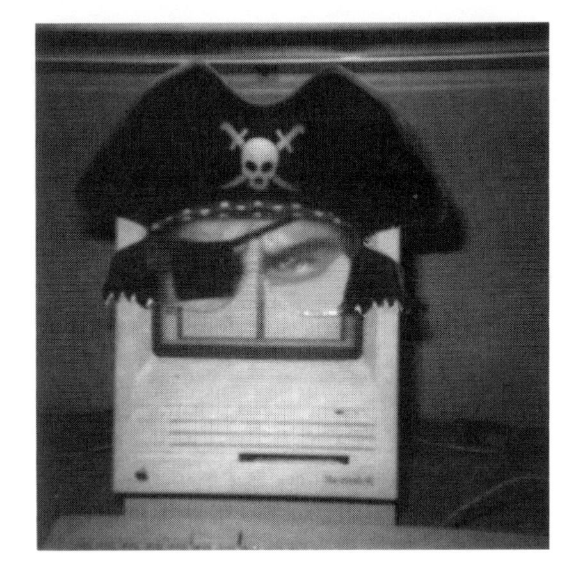

Figure 2

Get even.

I say it with a stare.

Get even.

I back it with a pistol.

Get even.

I say it because it is all I can say.

Get even.

Because I am programmed to do so.

Get even.

Because that's where the wires take me.

Even the elasticity of my plastic skin cannot alter the course,

as the boats in this ride cannot be altered.

All I can say, over and over and over again—until now,

now when an English course has freed me from the

warfare of wirefare—

I

Get Even.

I do not want to be there.

I am not a pirate.

I am not a skull-digger, or a looter like the South

Central pirates.

I simply get even.

This poem is my way out.

Now we're even.

Neal Modelevsky

Not everyone worked to capture in writing some of the qualities people ascribe to technology. That hardly mattered. What mattered a lot more was that the activity provided an occasion for interesting and evocative writing, writing that helped make strange things familiar and familiar things strange. The students were also introduced to the use of instant photography as a way to 'document' important parts of their work. (An anthology of pictures and writing was later distributed to participants.)

Get the picture

Here's a different sort of activity, good for teacher-training or for regular classroom use. You can either take pictures yourself or have participants take them. The idea is to create some interesting visual prompts for problem-solving activities, and it's definitely worthwhile discussing what it is about a picture that makes it a good prompt.

Here are two pictures, along with writing assignments that grew out of them.

1) If the person taking the step in this picture (Figure 3) is invisible, what will the person do tomorrow? How did the person become invisible? What was the person in this picture planning to do?

Figure 3

2) What is the cat writing about the objects surrounding it (Figure 4)? Find a reason for the 'cabin' of maple syrup to be included with this group of objects, which includes a miniature TV. What books would this cat be likely to read and why?

Figure 4

It's a great deal of fun (oh…and it's educational, too) to figure out what pictures to take, what assignments fit which audiences, and how to complete someone else's assignment. The instant-picture technology allows for fairly rapid testing of ideas and revision of visual prompts.

Dumb question/smart answer department

'How do you get kids to take such stunning pictures?'

This is the question I asked Joanne Koltnow, one of the pioneers in having students experiment with instant photography. Her answer? You just learn to recognize them when they come out of the camera.

Typical. I'd gotten so involved in technology and technique that I'd forgotten one of the most important bits of educational wisdom, sometimes you don't have to 'fix' people; you don't have to change them. You just have to get out of their way.

I relearned an even more sobering lesson from a book that Joanne lent me, *The Image is You*, edited by Donald Erceg, with text by Robert Coles (Boston: Houghton Mifflin Co., 1969). Unfortunately it's now out of print, but it's well worth searching for. In the 1960s, in an anti-poverty program in Boston's Roxbury ghetto, a group of children received some Polaroid cameras, a gift from the company to the Cooper Community Center. Walking though this area of the city, Erceg described it this way:

> I felt like the sole survivor of a war standing on an empty battlefield. I saw whole blocks flattened, vacant. The shells of buildings and stark trees stood over desolate piles of rubble. Standing in a playground with its bright-colored, wooden jungle gym and glass-littered asphalt, a row of broken plaster-of-Paris columns guarded the empty streets and fields. A pair of gray cats scavenging piles of rubbish and abandoned furniture across the street were the only signs of life.

Children at the community center used the cameras to photograph their world. As described by Erceg, '[the] children's photographs uncovered the moods of the neighborhood and its black community with disarming simplicity and frightening accuracy.' Psychologist and educator Robert Coles, in his commentary, offered this perspective:

> I hope that the photographs in this book and the children's words inspire admiration and surprise and awe and doubt and suspicion and skepticism. I hope that teachers and parents conclude that children have hope but need more chances, have thoughts and visions and fantasies and dreams but need more possibilities, need ways to translate what is experienced and felt and witnessed into what makes sense and works and does something, or is something. In the words of one child: 'First you look, then you go and click the camera; then you've got something. You can have it around you and you say to yourself, "Look at that, I did it".'

This book provides a striking example of how much we can learn from our students. It gives us the best kind of object-lesson in how important it is to give students access to both words and images, so they can better help us to see what they mean. In addition, Erceg and Coles remind us that when students are given the opportunity to document how they 'see' their world, they can provide a harsh reminder that we as educators need to re-acquaint ourselves with some of the broader contexts of the world in which our schools exist.

In *Teaching as a Conserving Activity*, media educator Neal Postman has pointed out that there are 'important consequences to changing the form of information, its quantity, speed, or direction'. Photography allows our students and us to play with all these factors, and the pay-offs can be extraordinary. It's a great way for people to show what they already know or are in the process of discovering.

A bigger picture

Written evaluations and follow-up conversations with Writing Project directors suggest that the NWP/PEP Alliance workshops are meeting their goals of providing valuable training, useful classroom materials, and significant opportunities for educators to meet in collegial, enjoyable, and professionally rewarding settings. The workshops are sometimes the centerpiece for a 'renewal' day; sometimes they're incorporated into Summer Institutes as a regularly scheduled session or as an optional 'enrichment' opportunity, one that is both challenging and entertaining.

From a theoretical perspective, there is ample support concerning the value in investigating and implementing our students' and our own capacities for integrating words and images. From a practical standpoint, photography, a widespread and familiar technology, offers an easily accessible and readily understandable medium for translating theory into practice. Teachers taking part in the TRIP project described earlier, and in the kinds of workshops described above, have demonstrated that there is a lot to learn and that they have a lot to contribute. In the effort to retain and expand the best of what we know about helping ourselves and our students both show *and* tell, we engage, in the fullest sense, in the language arts.

Visual information literacy includes the ability to 'read' (i.e. interpret, decode, translate) and design (i.e. create and communicate with) images. In addition, it implies an active rather than a passive orientation. We must decide what objects constitute meaningful features in our environment. We must decide what is worth looking for and looking at. The overall goal is to be able to see oneself in the world more clearly, more accurately, and more completely.

Note

1) The two selections that follow are taken from the 'Anthology of 1000 Words', as described above. In the first, Alice Slusher asks 'How can I possibly hope to reach [my students] without knowing who they are?' She discovers that her own roots help her answer important questions about her students and herself, and she tells a compelling story about a classroom in darkness and light. In a second account, Ginny Woodall-Gainey succeeds in bringing into focus a literal image and a metaphorical one, and in building a bridge between the two. By doing so, she discovers that seeing is believing and that believing is seeing. *–SM*

References

Argyris, C. and Schön, D.A. *Theory in Practice: Increasing Professional Effectiveness*, Jossey-Bass: San Francisco, 1974.

Armstrong, T. *Multiple Intelligences in the Classroom*, Association for Supervision and Curriculum Development: Alexandria, Virginia, 1994.

Boud, D., Keogh, R, and Walker, D. (eds.) *Reflection: Turning Experience into Learning*, Kogan Page: London, 1985.

Cochran-Smith, M. and Lytle, S.L. (eds.) *Inside Outside: Teacher Research and Knowledge*, Teachers College Press: New York, 1993.

Flower, L. Wallace, D.L., Norris, L., and Burnett, R.E. (eds.) *Making Thinking Visible: Writing, Collaborative Planning, and Classroom Inquiry*, National Council of Teachers of English: National Writing Project: Urbana, Illinois, 1994.

Gardner, H. *Frames of Mind: The Theory of Multiple Intelligences*, Basic Books: New York, 1983.

Grimmet, P.P. and Erickson, G.L. (eds.) *Reflection in Teacher Education*, Teachers College Press: New York, 1988.

Schön, D.A. *Educating the Reflective Practitioner*, Jossey-Bass: San Francisco 1987.

Schön, D.A. *The Reflective Practitioner*, Basic Books: New York 1983.

Smagorinsky, P. *Expressions: Multiple Intelligences in the English Class*, National Council of Teachers of English: Urbana, Illinois, 1991.

Smith, M.A. and Ylvisaker, M. (eds.) *Teachers' Voices: Portfolios in the Classroom*, Berkeley, California 1993.

Using the Computer at Home

Toni Downes

Fiona's Story

Fiona is 10 years old and she lives with her mum and dad and younger brother in their own home in an inner-city suburb of Sydney. Her everyday life involves the full round of normal family and school activities. Three years ago her mum and dad decided to purchase a family computer. Their primary purpose for purchasing the computer related to their own needs in their careers but they also considered the educational value for the children. With this in mind they purchased a relatively powerful desktop computer with a CD-Rom drive, a modem and a colour printer.

Both parents have a strong interest in the use of the computer, although Fiona's mum admits that since she has been working part-time her need to learn how to use it has decreased and consequently she hasn't used it as much as she had anticipated. However, she believes that access to e-mail has changed her life in the sense that she can now keep in daily contact with her sister and family in Canada. Fiona's dad has taken readily to the computer and particularly enjoys using the Internet for accessing a range of information for work and leisure related purposes.

Fiona and her brother are regular users of the computer. Her brother, who is 7 years of age, mainly uses it to play the very limited range of games that the family owns. Fiona has developed a much wider range of uses. Apart from game playing, she designs cards and posters, draws, writes narratives and factual texts, looks up information on the electronic encyclopaedias, and with the help of her dad searches the World Wide Web for information related to leisure and school-based projects. Both Fiona and her brother join in the family e-mail sessions with the cousins in Canada.

Fiona enjoys using the home computer and is very confident in her ability to use it. While she gets help from her parents from time to time, Fiona believes that she best learns to use the computer by fiddling around and discovering new features or ways of doing things. Fiona regards this type of activity as 'playing'. Recently she taught her mum and dad how to use the 'word art' feature within their word processor.

Figure 1

I love my cat

She discovered this feature herself, one day while she was 'playing' with the menus (Figure 1).

At school Fiona rarely uses the computer in her classroom. In fact, the classroom computer is hardly used at all and when it is the boys tend to take over. Fiona doesn't mind the boys taking over, as she is more interested in using the computer in her home. She explains her preference in terms of the computer at home being more powerful than the one in the classroom, and at home she can usually work in an uninterrupted fashion for as long as she needs. At school she is a library monitor and confidently uses and helps other students use the computerised catalogue and borrowing system several lunchtimes each week.

Over the last few weeks at school Fiona has been working on a group project about the Olympic Games. At home she used the search facility on the electronic encyclopaedia to look up information on the track and field events and on a number of the world's best athletes. While she was able to find some information from this source, her dad helped her search the World Wide Web for sites that had further information. All this information was printed out and taken to school for the children in her group to use. As well as using the computer for this research Fiona spent much time writing at the computer over these weeks. She wrote letters to three of her pen-pals, wrote e-mail to her cousins, wrote a narrative for a Writing Competition at the local library, wrote a poem about her grandmother (who had died recently) and composed an information report about one of Australia's track and field athletes who was taking part in the Olympics.

Fiona likes using the computer for looking up information and writing things as she finds it easier and makes her written work look better. She enjoys playing on computers because she can 'go crazy' on them: *' well... you just draw something really weird or something.... you can put it somewhere or change it or leave things...you can jumble it up or something and if you don't like it you can rub it all out.'*

If she could change anything at home Fiona would have more games where she could interact with people. If she could change anything at school she would have more computers *'so everyone could .....instead of doing work all day... you can play with your friends on the computer learning things like maths and playing with words, ...typing things in so your friends have to figure it out'.*

Fiona believes her life is different because she has a computer at home: *'.....especially on the net you can talk to other people around the world, it costs more to talk on the telephone because we have a flat rate*[1]*...... and we can see what's happening all over the world... instead of dad watching TV we can look up what's happening on the Internet and you can buy stuff on the Internet instead of going out shopping.'*

What research tells us about children's use of computers in their homes

Fiona is just one of the increasing number of children in today's primary classrooms who are confident, competent and regular users of computers in their homes. Her story has been reconstructed from two interviews and a journal kept over two weeks which recorded each of her computing activities. Over the last two years over 400 boys and girls have taken part in discussion groups, interviews and/or diary writing as part of a major research study on children's use of computers in their homes (Downes & Reddacliff, 1997; Downes, Reddacliff & Moont, 1995; Downes, Reddacliff & Moont, 1996). Each of these children had been identified by their parents as a regular user of a home computer. The children were aged between 5 and 12 and came from a variety of social, economic and cultural backgrounds in urban Sydney.

The purpose of this study was to discuss with these children the computer experiences they have within their homes and their schools, and their perceptions about a range of issues associated with their experiences. Significant themes that emerged from the study included issues of access, the range of computer use, the ways children learn to use them in their homes and children's perceptions of differences between computing at home and at school.

Access

The processes involved in selecting the children for the study clearly indicated that in all schools there were children who had no access to a home computer, children who had access to basic equipment (a computer and a printer), and a significant number of children who had access to a wide range of the latest technologies. These children tended to have parents who used computers in their place of work and lived in more wealthy school communities. The children in these families were twice as likely to have two or more computers in their home, and more than twice as likely to have printers and CD-Rom drives. Furthermore, these children's families owned almost all of the modems in the study.

The proportions of students in each group varied across the school communities. The Australian Bureau of Statistics (1996) now estimates that close on 50 per cent of urban households with children have a computer. In the more affluent communities in this study about 85 per cent of the children in the school had access to a computer in their home, while in the least affluent schools less than 20 per cent of children did. Research in Australia and other countries supports this notion of technology-rich and technology-poor families and communities (Australian Bureau of Statistics, 1996; Times Mirror Centre for the People and the Press, 1994).

For children who had computers in their homes, there were a range of factors which influenced children's access. These included location of the computer, ownership, who uses it most, and rules about its use.

Most families had sets of rules, either implicit or explicit, that impacted on children's access to the computers. The first group of rules, the community rules, included rules for managing the processes for shared use of the computer and its resources, resolving conflicts and defining acceptable behaviour. In general older family members have priority over younger ones, and people needing to do work have priority over those who want to play games.

The second group of rules, the personal rules, set conditions and limits on the children's use. These included when, how and for what the children could use it, when supervision or permission is needed, and codes of behaviour when using the computer. There were no restrictions on computer use associated with school work, but there were overall time and other restrictions on game playing. Many rules associated with game playing paralleled the types of rules surrounding television watching in the home (Cupitt & Stockbridge, 1996).

The existence of these rules indicate that families have clear views about the role of the computer as a leisure and work-related device and that they place priority on the use of the computer for work or school-related activity.

Generally both the boys and girls seemed content with the amount of access they had, although a number of boys spoke about restrictions on the amount of time that they could spend playing games. This contrasts sharply with the strong gender differences found in earlier research when computers in the home were more of an 'innovation' and had restricted uses such as game playing and programming (Wheelock, 1992). It can probably be best explained in terms of the parental expectations that 'work' on the computer has a higher priority than games, and the range of applications available on the home computer now includes wordprocessing, desktop publishing, drawing/ painting/ designing programmes and a wide range of multimedia encyclopaedia and factual texts. Both of these factors combine to increase girls' interest in and access to the computer.

The ways children learn to use computers in their homes

Most children learned to use the computer by having another person in the home, usually a family member, show or teach them what to do. Fathers and brothers were named as the tutor twice as often as mothers and sisters. Following the initial teaching period the learning was gradually handed over to the child who 'fiddled' and 'explored' and occasionally asked questions or sought help from a knowledgeable family member: *'my sister- because she knows more about computers than dad.'*

A small number of learners were taught by school teachers or extended family members such as uncles or cousins. Sometimes children learned by watching someone else use the computer, teaching themselves or by reading a computer manual or game guide.

If something goes wrong with the computer when it is being used about a third of the children were confident that they could fix it themselves. One older girl commented: *'Yes - just try things until they work'.* About a third of the children thought it depended on the problem, as an older boy explained: *'depends - if a minor problem I can, if its a disk I can delete things, but if its a big malfunction I call dad'.* The remaining third of the children thought they needed to get help in most situations. Interestingly gender had no relationship with the children's responses, yet older children were slightly more likely to have a positive response and children who had one or more parents using a computer at their work were almost twice as likely to have a positive response to the question. This finding suggests that the children in these families were being taught or were learning simple maintenance and problem-solving skills.

When children did need help, fathers were overwhelmingly the first person to be called. Uncles were also frequently called if dad could not help, for example a young boy said *'My dad calls my uncle, and my uncle, my dad and me and my friend fix/ do it together'.* Mothers and sisters were rarely mentioned by children as a person to be called when things went wrong.

The range of computer uses

The children used computers for a variety of purposes and were comfortable moving between playing games and doing work on the computer. While game playing remained the more common activity, many of these children regularly engaged in writing and drawing activities and used information-based programs for leisure as well as school-related work.

Game playing

Boys and girls reported the same frequency of game playing on the computer and the same interest in playing games. Cunningham (1994) argued that it was the shift from game playing in arcades to playing in the home, that enabled girls greater participation in game playing. However, in this study, gender differences were still evident in that boys were more likely to own and operate a dedicated video-game machine in the home and they tended to play different types of games to the girls (as did their fathers to their mothers).

Much has been written about children's game playing in terms of its effects. The popular press echoes the 'moral panic' approach often associated with children's watching of television. A recent review of the literature (Durkin, 1995) on the effects of computer games on young people found that the stronger negative claims are not supported by the growing body of research in this area, stating that: 'Computer games have not led to the development of a generation of isolated, antisocial, compulsive computer users with strong propensities for aggression.' (op. cit. p. 71).

Table I Frequency of Game Playing on the computer

Gender	Never	Hardly ever	Less than once a week	About once a week	Two or three times a week	At least once a day	Several hours
Girls	3%	1%	8%	28%	31%	23%	6%
Boys	-	1%	10%	24%	39%	22%	5%
All	1.5%	1%	9%	26%	35%	22.5%	5%

Based on the children interviewed in Stage 2 of the study: 275 children - 134 boys and 141 girls equally distributed over grades 3-6. (Downes et al., 1996)

In fact Durkin points to some research which indicates there may be some gains in cognitive and perceptual-motor skills through game playing. Certainly most children in our study believed that there were benefits from playing games. These included strategies specifically related to a game, to game playing in general or to general computing skills and knowledge. More interestingly some children reported that they learnt a range of 'real world' skills from the simulated environments within the games. Examples of these included physical skills such as driving cars, skiing, playing golf, playing soccer, controlling aircraft, shooting missiles, how to use guns, and other more strategic skills such as handling money, building cities and playing real card games. A small number of children reported that they learned how to solve problems, develop thinking skills, have patience and develop perseverance, memory and imagination.

Children's ways of describing their other uses of the computer

The children's rich descriptions of how they used computers were permeated with the language of 'play'. Regardless of what task they were undertaking, children, particularly the younger ones, used the word 'play' to describe what they were doing or the word 'game' to describe particular software. For example, when talking about their favourite game, a number of children mentioned KID PIX or other types of programs that enabled children to draw or paint. Some of the named software, particularly software designed for early childhood did indeed have 'game' contexts embedded within them, others however did not. Generally what this other software provided for the child was the flexibility to decide whether to use the software in a playful or purposeful way. One child explained: *'I can play*

PAINTBRUSH and print my favourite pictures'. Even the use of word processors, which allowed the children to create many different types of visual effects through manipulating font type, size, colour or shape (word art), and other features was sometimes described as 'playing'.

One younger child also carried this language over to his use of electronic stories: *'..I play books...'* and several older children spoke of playing with the electronic encyclopaedia and the information within it. In each case, when probed the children were referring to their control over how they used the software and/or features within the software which allowed them to interact with the objects or information within the software. One parent from Stage 3 of the study, who constantly used the computer in a professional capacity, also volunteered a comment on computer use and play from her own perspective: *'Sometimes I enjoy what I'm doing so much that it might as well be play'*.

While it needs to be recognised that there are many definitions of the word 'play', and that the term can be used to describe many different processes and activities, its abundant use points to the way that, at least in the minds of children, some aspects of computer use afford the user the same sense of enjoyment and/or control which is similar to that found in more traditional play activities and that this affordance extends to both work and leisure-related activities.

Using the computer for leisure-related activities other than game playing

Many of the children regularly write and draw and use information-based programs for leisure (as well as school-related work). They used the computer to write for a variety of purposes and audiences. These included stories, letters, party invitations, signs, poems, songs, cards, jokes, memos/messages, lists, captions, diaries/journals and recipes. Girls were more likely to report that they wrote for leisure on the computer, while younger boys (and children without printers) were less likely to write for leisure.

Some children wrote straight on the computer, others wrote on paper first and some children did both, depending on the task. There appeared to be a school / home difference in the minds of some of the children: *'Depends if it's for fun I write straight on the computer. If it's for school then I want it to be perfect, so I write on paper first'*. Some children preferred the computer because of the ease of editing: *'straight on the computer, it's much more fun, you can rub it out again'*. These differences were not related to age or gender.

Most of the children also indicated that they used the computer for drawing or working with graphics. The activities included: experimenting/playing within painting and drawing programs; making cards, banners, signs, posters and invitations; drawing pictures or using ready made graphics to illustrate stories, assignments or projects; designing things such as flags, dresses, houses and cities. Several children commented that they did not ever save or print out their work: *'I use Paintbrush to practise drawing...don't print it because we have no colour printer....*

Use the rubber to take it off the screen'; while some mentioned 'cutting' and 'pasting' their pictures into their writing documents.

Doing school work on the home computer

Many of the children interviewed described their home computer as a useful tool for doing school work. The school work they referred to included projects, assignments, essays, book reviews, research, making notes, answering questions, getting pictures, using the calculator for maths, practising times tables, making newspapers, finding the meanings of words, poetry, writing, spelling lists and speeches. Older students were more likely to report that they used the computer for school work, with 86 per cent of Year 6 children saying that they regularly did some kind of school work on the computer.

For the majority of students 'doing projects' was the main school activity supported by computer use. This involved accessing and/or presenting information. It usually involved moving between the screen and paper. Some children used screen and paper-based sources of information, others printed out the information from the screen, worked on it on paper, then typed the final version back into the computer. A small number of children worked within the screen environment until the final stages then wrote the final copy by hand. Most of these children did not have access to printers in their home. Some children commented that their teachers insisted that the final version be handwritten, others believed their teachers gave them a better grade because their work, produced on the computer, was well presented.

When obtaining information from CD-Rom encyclopaedias, many students were able to use their reading, summarising and editing skills. Many read through the information and selected information to put into their own words and retyped it back into the computer. Some students were able to do this within a fully screen-based process: *'go to the encyclopaedia, cut information from the encyclopaedia and paste it into the word processor. I change, add, delete and try to put it in my own words.'*

Others showed a lack of understanding of copyright and plagiarism as they literally pasted the print straight into their book or project or directly copied paper-based text into the computer. Two such children explained how they *'printed out the information. Read it. Cut it up with scissors because some of the information is too hard to understand. Pasted it onto cardboard'*; and *'went to the library - got book and photocopied it and just typed it up on the computer.'*

When referring to benefits of word processors in terms of their work, ease of editing, and 'the look' of the final copy were commonly mentioned. Children did not speak of ease of getting ideas down, of organising ideas nor of the quality of the content of their writing.

Another interesting feature of the children's comments about their school-related home computing was the number of children who mentioned the help they received with their work. Parents helped their children with many aspects of the work, such as printing out their work, in some cases typing it in, checking their work in terms of sentence structure, spelling and grammar and at times helping them put the text into their own words. Both mothers and fathers received equal mentions for giving this type of assistance. In many ways this assistance mirrors the more traditional assistance given when children are working with pen and paper.

Children's perceptions of differences between computing at home and at school

Frequency of computing varied in the home and school environments. While the majority of the children in this study used their home computer two or more times each week, few of them reported that they used the school's computers more than once a week. Many children indicated that they rarely used the school's computers at all, even if there was one readily accessible in their classroom.

Children who attended schools with timetabled sessions in computer laboratories reported more school use than their counterparts. A small number of children actually made critical comments about the lack of a system within their classroom for ensuring that all children had a turn at using the computer: *'Some people have had a turn but she always picks the same people'*. Particularly some younger children, both boys and girls, perceived that the computer was controlled by 'others' and they had been denied access. One child complained: *'People keep taking my turn - they just don't want me to get a turn. They keep playing it'*; another commented: *' When I play with Sarah she has all the turns, she is really bossy'*.

The overwhelming majority of children preferred to use a computer in the home environment rather than the school environment. The proportion of children preferring home use was greater in more affluent communities, where children had more computing equipment in their homes than in the less affluent areas. A smaller number enjoyed using a computer in either environment and did not want to select a preference, as a younger girl explained: *'There are lots of different games to choose from at home and at school'*. A few children preferred the school computers.

Many children explained their preferences in terms of the differences in the type of computer hardware and software available in the home and school environments. For some children frustration was caused by the limitations of some computing equipment: *'My computer (at home) stops in the middle of a game. School has a better computer, it works properly'*. Conversely some children were frustrated with the equipment at school: *'At school the printer sometimes doesn't work, so it's unreliable'*.

Some children had more sophisticated computing equipment at home than in their local schools. When older boys described the differences they often referred to the technical specifications of the computers: *'At home my computer is better and faster with more memory and programs. I have 500 megs of memory at home. The school computers only have 40 megs'*. In general the comments related to what you could do with the computers at home: *'My computer at home has better things on it, more information, more programs and it's more modern'*; and *'The ones (computers) at school don't give you all the information you want'*.

Other reasons that children preferred to use the computer at home were the quieter environment, their familiarity with the computer, less restricted access and control over and ownership of the computer. The children's comments ranged from *'because it's fun at home, you have more peace and quiet and no children talking'*; *'because I can use it anytime I want, not like at school'*; to *'I can do whatever I want instead of what the teacher says to do. I get more time to think'*. A number of girls made explicit comments about boys having more control than themselves at school: *'at home, I've got it to myself because at school all the boys hog it'*. A number of boys also recognised the competitive element at school: *'at home, you don't have to compete with others for a turn.'*

In the home and school environments children said that they were usually engaged in different computing activities. One older girl explained: *'At school, we do postcards, stamps, illustrations and designing. At home, I type projects and reports'*. In fact, typing projects and reports, which involved sustained periods of computer use, was mainly a home activity. Many children chose to take this type of work home, even when the project was a classroom rather than homework activity. A few specialised computing activities were enjoyed at school because they were not available in the majority of homes such as designing and making CDs, learning touch typing, scanning photographs and producing newspapers, making video clips, working with animations, using a modem and the Internet.

Marina's Story

A narrative of one teacher's attempts to further develop links between home and school computing

Marina has been teaching primary school children for the past 10 years. Four years ago she and her husband purchased a computer for the family home to help them with their work and also for their children who were in secondary school. Marina uses the home computer for writing letters and notes and preparing worksheets for the children in her class. Recently the family subscribed to the Internet, as the oldest child, Mark, was taking computer studies at school. Marina sat down with Mark on a number of occasions when he was accessing the Internet to see what all the fuss was about. She was much impressed by the French language discussion group they located one evening and by the range of information on topics related to the history and geography that he was studying at school.

At school, Marina has access to five computers which she shares with the two other Year 4 Classes. These computers are in the corridor outside her classroom. Two of these computers have CD-Rom drives and two have printers. There is one Internet connection, in the Library.

When Marina first began using computers with her students several years ago she mainly used topic-based software to support some of the outcomes of her integrated units of work and Logo to support some aspects of the Mathematics programme. These days she has expanded the classroom use to include word processing, desktop publishing and accessing information from electronic factual texts such as CD-Rom encyclopaedias. Over the last three years she has noted an increasing number of children in her class handing in work which has been word processed and bringing in pages of information which had been printed out from various electronic texts and databases. To her consternation, she has noticed the growing incidence of obvious plagiarism in word processed project work. While she has always had children in her classes that have copied texts directly from encyclopaedias and other texts, Marina wondered if it was easier with a computer and more tempting because of the sheer volume of information that the children were now accessing.

At the beginning of the school year she surveyed the children in her class and found that five children had computers, CD-Roms and printers (two colour), one of these also had a modem and an Internet subscription, four children had computers and printers, and two children had very old computers mainly used for games and a little word processing. Five other children mentioned that they occasionally used computers at their friend's places or the local library. The remainder of the class, thirteen children, did not use a computer outside of school. She also found that eight of the children were quite reasonable typists, two being exceptional, seven had reasonable familiarity with the keyboard while the remainder of the class still were at the 'hunt and peck' stage.

Based on this information and on her planned curriculum outcomes, Marina added the following items to her normal organisation and plans for computing in the classroom.

1) At each of her parent information sessions during the year Marina:

- explained the role of computers in her planned programme
- explained how computers in the school and the local council library were available at different times for children to use, if they did not have access to computers at home
- explained to parents how they could help their children better use the computers in the homes and the local library to support their school work
- recommended to families with computers some Australian encyclopaedia and factual texts on CD-Roms (to counteract the American bias in the more common CD texts)
- explained to parents her approach to developing all children's computing skills within her normal classroom programme.

2) A 'quick type' tutor program was added to the collection of software on the computers. During spare and free time children who were working their way up to level 5 of the program had priority over all other users for 10 minute intervals. After several weeks, this priority access was restricted to the two older machines.

3) When children were booked at the computers for writing, they worked in pairs for composing, with a 'thinker' and a 'typist'. Partnerships were changed on a regular basis so that sometimes children worked with a faster typist and other times they were the faster typist. A number of children also worked as peer tutors in writing groups with Year 1 and 2 children. In these groups they were the typist.

4) From time to time children in the class acted as computer monitors in junior classrooms. This usually occurred when the younger children were being introduced to a new topic-based piece of software. Children who did not have a computer at home were included in this programme but were offered explicit instruction on any unfamiliar loading, playing, saving and printing routines. This training was carried out by other students where possible.

5) In group-based project work, Marina allocated particular children to gather information from different sources, such as the school library, local library, CD-Rom encyclopaedias (home, school or local library), the Internet (at home, school or local library), and special people and organisations, as appropriate. Based on a well-structured set of questions, each group member scanned their source for information relevant to each question. The children added their found information for each question to a group chart. In pairs, children would construct answers to each of the questions drawing on all the information gathered. In addition a progressive series of debriefing sessions were held with all children about:

   - differences in the nature of the information from the different sources
   - which sources were better suited to different types of questions
   - what information was easier to find in what sources
   - the qualities of the information found, e.g. readability, currency, authority.

   These discussions led to advice that children should always use a variety of sources – electronic, printed and human – for their research.

6) Marina asked the teacher librarian to add the following skills to the range of research skills she would teach the class over the first term:
   - searching electronic texts, topic based and 'free text' searching and how to narrow and widen searches
   - referencing electronic sources as well as books, referencing images and pictures as well as text-based information.
7) Marina modelled the processes of finding information in an electronic source, 'thinking aloud' about the information in terms of the questions/tasks and the construction of her own sentences to represent the relevant information for the questions/task. She also set clear guidelines about what were acceptable practices in preparing project presentations regardless of the source of information.
8) She introduced the whole class to 'electronic slide shows' as one medium for presenting project work. She encouraged children to use this approach using either their computers at home or those in the school library. As part of this process all children had 'booked' computer time to work with a partner to develop skills in transferring pictures from clip-art and photo libraries into their own slides.

Discussion of the classroom implications of increasing home use of computers

Equity

Researchers investigating computer-integrated classroom settings are increasingly aware that computers can promote and even augment inequalities among children. The research further indicates that income, gender, race, ethnicity and language are factors that may preclude minority groups from deriving the type and extent of access, participation and benefit enjoyed by their middle class English-proficient counterparts (DeVillar & Faltis, 1991).

These inequalities are being compounded as increasing numbers of children from more affluent homes gain further benefits from their access to and use of home computers. Exacerbating the equity issue is the likelihood that children who have no access or only minimal access to basic equipment in their homes are also likely to come from families who do not share equally in the social and workplace benefits of computing technologies, nor have access to a range of other literacy processes and artefacts in the home.

These inequalities are complex and plans to redress them need to be well considered. De Villar and Faltis (1991) found that the factors relating to differences in participation and benefit were not so much pertaining to the technology or children's confidence and competence with the technology, but to the general classroom culture, group dynamics, and the way teaching and learning was organised. These factors are often also quoted in research focusing on gender inequalities with technology in schools. In the Downes study (1995, 1996 & 1997) many children were critical of the organisation and management of classroom use of computers, including situations where boys were allowed to 'hog' the computers. Interestingly, even some very competent and confident home-computing-using girls opted out of using computers in their classrooms because of the competition.

In addressing these issues teachers need to:

- create a classroom culture where all children can participate in and benefit from all teaching and learning activities, including those using new technologies
- develop children's attitudes, values and skills which enable all to work both independently and cooperatively with their peers, sharing resources as needed
- organise teaching and learning experiences in ways that ensure all children have access to and use of necessary resources and can participate in all aspects of the task. For example, when establishing pairs or groups, one criteria could be placing children of equal assertiveness together. A key feature of class groupings should be that groups/pairs vary over time, and that all children have the opportunity to be leaders as well as team members, and experts as well as novices within the task domain. This approach reduces the incidence of group members becoming reliant on the complementary expertise of fellow members in long term groupings, as found in the research of Hoyles and Sutherland (1989) on maths and logo
- allocate time, particularly for children who do not have access to computers outside of the classroom, to explore the use of the computer or engage in purposeful tasks on their own. This will include some explicit teaching of processes and the opportunity to practise new skills. Peer tutors, who have expertise can coach less knowledgeable classmates, and in turn these children can pass on new-found skills to others, including children in earlier grades or parents who might visit or assist in the classroom from time to time
- design teaching and learning activities and management regimes which maximise children's sense of control over the use of the school's computers. Strategies need to move beyond booked computer time to take account of the timely access to computers at the point of need, possibly for extended periods of time. Again these issues are not isolated to use of technologies but are more broadly linked to managing the use of any scarce resources, or in the context of technology-rich schools, training children to accept more responsibility for managing their own learning time

- create and nurture links between school and home. The initial step would involve finding out what resources are available in the children's home. Strategies for strengthening links could include: inviting parents into classrooms to observe how children learn with computers, having information nights about computers for parents, providing parents with information about community computing courses (for themselves), places where children can use computers in the local community if they don't have access at home, for example the local library, and if school resources permit, having a set of laptop computers that families can borrow for periods of time.

Gaining and losing skills

In general parents and most teachers (Downes & Reddacliff, 1997) strongly believe that new skills are essential to survival in the wider world where electronic texts are increasingly the major source of social information, argument and entertainment in our society and that schools need to take leadership in this area. At the same time they spoke at length of concerns about children losing some existing skills and understanding associated with accessing information and communicating within the world of print. Concerns about poor handwriting, spelling, reading books and using a library were common. Many want today's children to straddle both technologies and develop expertise in both. One mother shared her frustration about her son who virtually puts all his writing on the computer *"I say 'Why are you using the computer like this? I want you to write'. I say 'You'll forget how to write soon if we don't stop this."* (Downes & Reddacliff, 1997). Many children, however, as they increasingly access a wide range of electronic tools and resources from their homes, are expressing a growing disinterest and /or frustration in print-based processes and technologies such as handwriting, using books as a source of information and using libraries: *'it (the Internet) is not like a library where there is a limited number of books and where everyone borrows the book on the subject and there are no books left'* (Downes & Reddacliff, 1997).

While these comments clearly demonstrate frustration on the part of the parents as well as the children, they also indicate the dilemma faced by children in today's world. While their parents and teachers can understand and articulate the need for children to straddle both worlds, many children are not only losing some paper-based skills but also the motivation to be bi-literate inside classrooms, where literacy is almost exclusively defined in terms of paper-based technologies. Scant attention is paid by home or school to the systematic development of control over the processes and understanding needed to effectively communicate and handle information with the new technologies.

In addressing these issues teachers need to:

- work with a broader range of 'texts' and media within their literacy programmes, including spoken, written, viewed, performed and interactive texts for local and remote, personal and mass audiences, using both traditional and electronic media. Such work includes the use of technologies beyond 'the book' and 'the computer' to include telephones, faxes, audio tapes, video-tapes, broadcast television, the Internet and other computer networks
- achieve a balance in terms of priority and effort to help children develop skills and understanding in all forms of texts, and help children develop a critical awareness of the appropriateness of text type and media to purpose and audience
- move beyond the use of the computer as a word processor of 'text to be printed', and include the computer as a maker of electronic texts such as data bases, hypertexts, multimedia presentations and World Wide Web pages and documents
- develop reading and research skills within both print-based and electronic media, stressing issues related to copyright and plagiarism and the important skills of skimming, reading for meaning, summarising and integrating information from a variety of sources
- develop children's visual literacy skills, including an understanding that images, just like texts, are constructed for a particular purpose and audience. They need to work with the notion of 'reading' images in both print and electronic form, while developing an awareness of the electronic media's reliance on the constructed image and less elaborate spoken and written text to convey information
- integrate a wide range of traditional and electronic resources into a variety of teaching and learning activities within the classroom.

A range of classroom examples and practical suggestions for integrating these strategies into primary classrooms can be found in *Learning in an Electronic World* (Downes, 1995).

Conclusion

It is important that teachers do not ignore the 'equity' issues surrounding the differential access to computers and related resources in homes. It is equally important that teachers do not use the equity issue to argue against recognising and capitalising on the skills and understanding that some children bring to school because they use computers in their homes. With reflection, it is difficult to justify a position based on a reverse equity argument that children with computers in their homes would have an unfair advantage. The fallacy of this argument is obvious if an analogy is drawn with print-based literacy processes and artefacts. Would any teacher deny children access to books and writing implements in the home, or ignore children who come to school already being able to read? Teachers need to develop strategies and processes so that all children within their classroom benefit from the extra skills and understanding that some children bring to class. As well, they need to consider a range of strategies for developing the skills and understanding of children who do not have access at home. Similarly when setting homework and project work, teachers need to help children effectively exploit all the resources available to them at home and at school. Encouraging children to use a variety of sources, both print and electronic, and structuring teaching and learning experiences to improve children's information handling skills across all media and information types (images as well as words) is a sound beginning. This approach will better equip today's children to be bi-literate, able to effectively communicate and handle information in both the world of print and the world of electronic media.

Notes

1. In Australia local calls are charged at a flat rate – they are not timed. As well, many Internet subscriptions are also flat rate.

References

Apple Computer Australia Pty Ltd. *The Impact of Computers on Australian Home Life* (1). Sydney: Apple Computer Australia, 1996.

Australian Bureau of Statistics. *Household use of information technology.* Canberra: Australian Government Publishing Service, 1994.

Australian Bureau of Statistics. 'Household Use of Information Technology', *Australia Catalogue* **8128.0**. Commonwealth of Australia: Canberra, 1996.

Cunningham, H. 'Gender and computer games'. *Media Education Journal* 17: 13-15 (1994).

Cupitt, M., & Stockbridge, S. *Families and Electronic Entertainment.* Australian

Broadcasting Authority and the Office of Film and Literature Classification: Sydney, 1996.

Downes, T. *Learning in an Electronic World.* Primary English Teaching Association, Newton, NSW, 1995.

Downes, T., & Reddacliff, C. *Stage 3 Preliminary Report of Children's Use of Electronic Technologies in the Home.* University of Western Sydney, Macarthur, Campbelltown, NSW 2560, 1997.

Downes, T., Reddacliff, C., & Moont, S. *A Preliminary Report of Children's Use of Electronic Technologies in the Home.* University of Western Sydney, Macarthur, Campbelltown, NSW 2560, 1995.

Downes, T., Reddacliff, C., & Moont, S. *Stage 2 Preliminary Report of Children's Use of Electronic Technologies in the Home.* University of Western Sydney, Macarthur, Campbelltown, NSW 2560, 1996.

Durkin, K. *Computer Games. Their Effects on young people. A Review.* Office of Film and literature Classification, Sydney, 1995.

Hoyles, C. & Sutherlend, R. *LOGO Mathematics in the Classroom.* Routledge: London, 1989.

Times Mirror Centre for the People and the Press. *Technology in the American Household*: Times Mirror Centre for the People and the Press, 1994.

Wheelock, J. 'Personal computers, gender and an institutional model of the household'. In R. Silverstone & E. Hirsch (eds) *Consuming technologies: media and information in domestic places,* (1992) 97-112. London: Routledge.

The Place of Learning

Moira Monteith

Introduction

The child is central to her or his own learning whatever model of teaching and learning we subscribe to. Children move between school and home yet often seem to want to keep some sort of independent control between the two. Children have experiences at school which they never tell their parents about and these experiences include formal learning situations. Often there are direct links between the family and school but these are secondary compared with the pupils' direct experience of both. Computers (to use the term generally employed by teachers and families) shift position, not only within this research project in going between school and home but dynamically in social terms.

Computers in education began as entirely separate items found in some classrooms and some homes. They could now, if that is what we require, take over the function of the locus of learning in the formal educational sense. So learning would then take place where the computer is. It is likely that such a situation would give extra 'control' to the learner, not merely by using portable computers in whatever form becomes available but because the oral context changes from school to home. Research with three Sheffield schools has shown that children speak frequently when working with parents and other family members. This confirms the findings of the Bristol Study (Wells, 1986) where Gordon Wells and his team found that children had far more opportunities to talk at home compared with school. They also found that 20 per cent of the dialogue children had with teachers came via questions and that most of these questions were 'display' questions where the teacher is asking the child to reproduce knowledge the teacher already knows. The dialogue at home, when children are word processing, appears rather different, with parents asking questions where they are actually seeking an answer rather than a confirmation. Children appear to have more control in terms of what they answer. They often ask questions themselves and the dialogue appears to indicate a different kind of relationship to learning and the task in hand than would be found in most classrooms. Therefore home learning could offer more independence for the learner.

A project such as *Children and parents word processing at home* can be seen in one sense as 'on the edge' of school work in that it is not concerned with the National

Curriculum as such. Its significance resides in the place of research, the interface between home and school learning. It examined what parents and children talked about as they worked together at word processing at home. It can be also a helpful illustration of a small research project carried out within a few schools where teachers and university staff can work together to evaluate certain learning situations or lines of action. Participating teachers can use such material in writing up a Masters-level dissertation or equivalent. This happened with a previous project involving portables (part of the NCET Portables Project) in one of the three schools.

Schools and Parents

English schools of necessity are involved in close co-operation with parents in a number of ways. Indeed, this requirement was specifically spelled out in the *OFSTED Guidance on the Inspection of Nursery and Primary Schools* as published in 1995, p 96: **Inspection Schedule 5.5: Partnership with parents and the community.**

Inspectors must evaluate and report on how links with parents contribute to pupils' learning and how the school's work is enriched by links with the community....Overall judgements need to establish whether:

- there are clear lines of communication
- the school's approach to relations with parents is maintained consistently
- the school does all it can to gain the involvement of all parents.

The growing number of computers in the classroom, at home and in business, has had a cumulative effect on all of us so it is not surprising that many parents and guardians want to know what kind of computer and software to buy for their children to use. Organisations such as PIN (Parents' Information Network) have flourished in the last year or so, catering particularly for the families' need for information. Clearly the previous routes for liaison between families and formal education such as the Local Authority Education Services, the National Council for Educational Technology (NCET) and generalised parent groups did not meet the need for information focused on home use of computers. The rapid growth of PIN indicates that parents not only wish to know what is happening with regard to computers and education but to act upon that knowledge. Families now assume that learning will go on at home with the computer. Even more, many families (according to PIN reports and interviews with parents in this project) believe that a home computer is becoming a necessity in order to help advance their children's learning.

Parental interest in the use of computers at home and in school coincides with a national endeavour to encourage parents to become more involved in and even responsible for their children's learning. It appears from Government initiatives and published papers, for example the White Paper *Excellence in Schools* that more homework is to be the norm. Indeed the Secretary of State for Education, David Blunkett said explicitly in an article in *The Guardian* (July 15th 1997):

> Parents are a child's first teacher. They need better information and advice to increase their involvement in their own child's learning. Home-school agreements will set out rights and responsibilities of home and school, explaining clearly what is expected of the school, of the parent and of the pupil.
>
> Such agreements will make clear the need for regular and punctual attendance, for good discipline and for the vital role which homework can play in supporting learning in and out of school.

Not that the move towards more homework has a particular connection with computers. Education systems such as that in Ireland have expected considerably higher levels of homework for some years compared with England. However, the national push for parental involvement in learning via OFSTED and also the Department for Education and Employment (DFEE) plus the parallel parental interest in computers fortunately coincides with the motivational effect computers bring to learning. The use of information communication technology brings with it a strong motivating force, revealed for example at the Chapel Town and Harehills Computer School in Margaret Cox's publication (1997) on students' motivation. This nexus of interests is the context in which the *Children and parents word processing at home* project took place.

Learning Models

If schools follow the OFSTED recommendations (1995) and hope to involve *all* parents as partners in their children's education, what kind of learning models can they offer when it comes to the use of computers? In a previous study (Merchant and Monteith, 1997) we found that parents followed the model for writing as practised by the school their children attended. So, for example, if a school had a policy promoting emergent writing the parents followed the stages of emergent writing while word processing with their children. However, if Rupert Wegerif is correct in believing that 'Computers are not always integrated effectively into the primary curriculum partly because teachers do not know how to conceptualise their educational role' (Wegerif, 1995) then the model for computer use may not be clear-cut and may well differ from teacher to teacher, even though schools have a defined ICT policy. Indeed, one of the two OFSTED recommendations specifically relating to primary schools and their use of ICT stated: 'The great variation in teaching IT capability in different classrooms in a primary school needs to be reduced'.

In addition, parents may well bring their own experiences to bear on their concept of what learning is. Children intuitively seem to 'get on' with computers.

At least that is a fairly general conclusion but possibly learning in this area is more a matter of what older people have to jettison, in order to learn in the same way as most children approach work with computers. Wendy Newcombe, a lecturer in a FE College explained this very effectively in December 1997: 'I believe fear is what prevented me from starting with computers. I was afraid I would break something expensive, afraid I would not be able to learn all the jargon, that I would make a fool of myself in front of colleagues and students who all appeared very knowledgeable in all aspects of computers.' Most of the adults my colleagues and I have been involved with in their early stages of computer use bring this set of attitudes with them.

We know that not many children have turned out to be prodigies in the past but a multitude of children in many families now are considered 'expert' at using the computer. Many of them, I suspect, achieved this 'status' without imitating the model of learning how to use a computer as exhibited by their parents. One of the parents in this project clearly left the 'computer side' of it to her daughter. When discussing printing out with another parent she said, 'Rachel will show Arlene 'cause I'm not..err, I'm hopeless without Rachel showing me.'

If we devise a learning equation for such a complex whole we inevitably include a number of variables which make any results difficult to assess except in a qualitative approach. For example:

school's models of learning + school's attitude(s) to ICT + parents' own learning experiences + discourse + child's attitude towards computers and habits of learning	= ?

Brief Description of Project

The three participating schools already had strong links with Sheffield Hallam University. They joined with other schools in the original NCET Portables Project and continue to be partnership schools taking teacher training students from the university. Two of the schools are in large housing estates dating back to the fifties or earlier, one is in a multi-cultural area with mainly older houses privately rented. One school is Nursery/Infant another Primary and the third a Junior school. As in many such Higher Education/school projects a team approach is very important (Underwood and Monteith, 1998). Francis Howlett from the NCET, participating staff from schools and university met together to decide on procedures. The schools were enthusiastic to manage the project themselves, so they chose the families with whom they wished to work and organised the approach to parents. Each school had between six and ten laptop computers with simple word processor, spellchecker and thesaurus, which they lent to families for up to a term at a time. The only stipulations the university made were that the parents taped their conversation as they worked with their children and the school kept a record of who used the computers, and wrote a brief report at the end of the project. University staff contacted the schools frequently and visited the school each time there was a 'new' set of parents. Small audio recorders were allocated to each family plus tapes. A number of parents were interviewed during the project.

Some 47 pupils with parents participated fully with another six who gave up owing to illness or other personal problems. University staff listened to all the tapes and 13 tapes were fully transcribed. The majority of parents were mothers although four fathers took part in the taped conversations and several more did attend meetings about the project at the three schools.

The project planning followed the theories of learning which bring together individual learning development, both in terms of competence and cognition, and its responsiveness to social and cultural pressures as exemplified in work by Jerome Bruner (1971), Neil Mercer (1995) and Gordon Wells (1986). The project team felt that such an approach is particularly appropriate when investigating the use of computers by individual pupils, with parental input and school influence.

The following analysis examines predominantly the talk between parent and child partnerships in four cases; a consideration of tapes from:

(a) two pairs indicating differing school approaches
(b) four children with the best collections of printed work
(c) pairs where the parents clearly had to persuade their children to take part
(d) two children who were mainly left to work on their own

The analysis takes as its starting point the research by David Wood (1992) when he was looking at 'teacher talk'. He mapped their talk in terms of 5 'conversation move types':

1 Enforced repetition	Say "I have one at home".
2 Two-choice question	Did you have a good time?
3 Wh-type question	Where did you go yesterday?
4 Personal contribution	I think sugar is bad for you.
5 Phatic Oh, lovely.	

He found that the teachers who used most of the first three move types had pupils who gave short responses, hardly ever elaborated on any answer to a question addressed to them and seldom volunteered any of their own thoughts or ideas even when they had the chance. Pupils in such classes also appeared to show signs of confusion. On the other hand, he found that teachers who 'speculate, suggest or surmise, inform, interpret or illustrate or simply listen....produce the mirror image of this pattern'.

Case A

In a comparison of parents' talk with children who are just beginning to read and write our previous finding (Merchant and Monteith, 1997) was confirmed. Parents tend to follow the model of learning provided by the school. Working with a school policy which encourages an emergent writing approach, parents tend to take note of their child's progress. They comment on improvement in the recognition of letter and computer keys (eg caps lock, space bar and delete). Elaine's parent includes comments each week: 'Went through the alphabet, names also easier to do. Now knows the sounds of letters but has difficulty locating them on the keyboard. ...... Elaine is now very familiar with the delete button and now appears to make a mistake, identify and correct it.....Gave her 3 letter words to do. I then left the room for her to do them by herself after which we corrected them. It took awhile. I don't think she has the confidence in her own ability to identify the sounds.' Parents comment on independent writing: 'starred work is Elaine working on her own'. Another child, Joanne, wrote the 'first nine words herself'. Parents accept mispellings as part of the learning process:

I can dror mum

I Can dror Jamie.

My dad's nim is Crag.

I hav got a dog.

His nim is Pepe.

I like riting.

I like woching cartuns.

However, they suggest making space between words, putting in capitals and fullstops. Children's work includes a wide variety of material, from 'I spy' games, nursery rhymes, sums, to 21 football teams with sponsors, and favourite TV programmes.

A second school has also suggested ideas to the parents as to what they might like to encourage the children to write about. The school does not have an emergent writing policy and possibly has not communicated entirely successfully to parents its strategies as regards learning in the early years. Emily's mother uses her own knowledge about language to encourage her daughter but this strategy is rather too early in Emily's writing and listening development. As a mother, she is probably also remembering her own experiences of school and so the whole tape is a series of short questions and answers. Many of the questions are similar in nature to those described in Woods' research, as being used to elucidate answers already known to the questioner. Parent and child go through all the colours of the rainbow, colour by colour and letter by letter. Emily has trouble with the word 'purple'. First she suggests 'p p' as the spelling. Her mother asks her if she knows what vowels are. Eventually by dint of her mother's helpful questions Emily works out that there are five vowels and lists them. She then has trouble spelling 'blue'.

Mother: what are they called? I've told you.

Emily: Err ...files.

Mother: (loudly) vowels

Emily: Fi-owls

Mother: vowels

Emily: files

Mother: no, you're saying it wrong. Vowels.

Emily: Fi-owls

Mother: that's it. Right. Vowels.

The conjunction of two abstract concepts, files and vowels, does not help Emily's understanding. Her mother organises the dialogue by asking questions and hinting at the answers so that Emily finds herself having to spell words such as 'reindeer' and 'sleigh' both of which are quite difficult for someone of her age. Emily has a go at 'sleigh'. She suggests 'o', 'u', 'i' in turn for the letter after 'sl'. In fact, Emily does know about vowels although she doesn't always get the correct one or even necessarily put one in. She is at the stage of realising that there are other letters than consonants when she is writing a word, and when her mother asks for the next letter suggests a vowel but not always the correct one in context. It would be helpful if parents knew that word processing is particularly useful for revealing stages of awareness about language (Monteith, 1996). Emily could be praised for what she is doing rather than made to feel she has made a mistake.

Emily does make at least one attempt to take over the process:

Emily : Right I'm going to read this... Red and yellow...

Mother: No, we've already read it, haven't we? You don't need to read that again. No.

Emily: Why?

Her mother then repeats an earlier question and they are back to the old pattern. At times she tells Emily to 'sit up' and to 'concentrate'. Since her tone is warm and pleasant and seems at variance with her comments and controlling list of questions, it seems likely that she has fallen back on the model of classroom interaction which she remembers. Later on she makes a comment to the listener: 'Emily hasn't been on the laptop computer for a few weeks because she was poorly with a very bad cough and a chest infection. And the other weeks where she hasn't done any work, is because she didn't, really, know what to do on it...' They then continue in the same manner as before.

Case B

The tapes accompanying the four best collections of print outs from one school have some similarities. The criteria for selection of 'best collections' were length and number of individual pieces, variety of topics included, use of paragraphing, and quality of prose. The last item is of course very subjective, but the work of the four pupils here was markedly better than the printed out selections from other pupils.

Three of the collections are by girls and one by a boy. The boy, Nisar, was in Year 6 and wrote at length, focusing in particular on cricket and a trip to Pakistan, which was movingly described. The very last item in the printout reads: 'I liked doing my laptop computer it is a really good project. It is a very good and helpful computer. My mum doesn't know anything about computer's (sic) but she has been helping me and she thought it was very easy and simple. She enjoyed working with me.' Unfortunately Nisar's mother's voice was never recorded. Instead we hear Nisar working with his tutor. Nisar's diary account was very detailed, not only about the runs scored. He divided the day by giving the times of his activities as well as describing what he did. 'After that I went to Stephens house I came back at 5:00 then I had a slice of swiss roll before my tutor came. He comes at 6:00 and goes at 7:15...' Another entry indicated that he worked with his tutor on the lap-top. The tape indicates that Nisar was very much in charge of what he was writing. The tutor gave instructions while Nisar typed steadily. 'Don't have too many ands in a sentence... ' and suggested that the 'best thing is to describe the game, make it exciting. Before the afternoon, full stop..... Doesn't make a complete sentence. Doesn't make sense.' He asked Nisar to read it out, and repeated that it had to make sense. Nisar did read it out aloud when requested but did not respond very much to instructions. Nevertheless his work was well constructed and corrected. Occasionally he replied to questions as to whether he was a bowler or a batsman, and which cricket side he was on. The word processing fitted into a very organised day and was given its due time. However, the tape indicates very little dialogue as such, certainly no partnership between parent and child as is represented on other tapes. According to a teacher, Nisar's mother went through his print-outs and looked for spelling mistakes. 'She wants him to do well.' It may be that his mother not only corrected the final version but helped him in discussion about his trip to Pakistan. On a tape from another pair, where the daughter wrote about her first trip to the West Indies, the mother took the lead when they were talking about the holiday. She knew the names of the places, reminded her daughter of details, and was particularly knowledgeable about the kinds of trees growing there.

Noreen's mother 'began' one session on tape by saying *'Right...Carry on with the story...Are you going to read what you've done before?'*

She then asked questions to which her daughter responded as 'the writer'.

Mother: What's he going to do?

Noreen: He's going to tell the deer to, um, go and get in where the hunters are based. And she runs and hides, because she knows her glen, when hunters are around and see her. And then he can get time to run and move the rabbits.

Mother: Yeah. What would happen if the hunter shot the deer?

Noreen: Well they won't.

Mother: They won't. OK.

Noreen: It will just be part of the story that they won't.

Mother: OK. Yep, sounds fine. So the squirrel wants the deer to distract the hunters.

Noreen: Yeah

The questions in this extract clearly were there to enable the mother's understanding. They confer the position of 'writer' on Noreen, who had the answers as regards her plot. However, Noreen, a Year 4 pupil, found the questioning helpful as later on she used the word 'distract' which possibly she would not have done had her mother not used it in her 'summing up'.

This pair also spent considerable time talking about spelling and punctuation. Noreen's mother asked her at the beginning of the tape: ' Do you want me to tell you if the spelling's not quite right?' and just after that, when the machine made a noise, 'Now that is a beep telling you that the spelling's not quite right.' Noreen made no response either time. Later on, when the screen must have looked wrong in some way, Noreen said: 'It's on the wrong line. Is that right?' They then worked together to organise the text more effectively.

Mother: You need a capital to start with, as it's a name...Now think about it...Where's the big A going to end up? Can you get yourself out of that one? What do you need to do? ...You want a space there.

Noreen: it doesn't even know how to spell Mummys.

Mother : well, put an apostrophe between the y and the s and see if it likes that.

Noreen: right...which is apostrophe...

Mother: let me see...Apostrophe is that one. Well done! well spelt!

Arlene, a Year 5 pupil, and her mother worked together through the instructions for setting up a new file. Arlene, possibly the most accomplished writer, was enthusiastic from the very beginning, asking how long she could type for and could she write what she wanted. She then asked questions about using the shift key and deleting. Her mother added a comment about the wrap around text. She asked 'Do you want me to tell you when you've spelt something wrong?' Although Arlene said yes to that question, some time after she stated: 'Can I just do this please? I don't like commenting on.' They worked together discussing rhymes but Arlene knew what she wanted to achieve. '...the leaves of the trees have different colours...No, make it into a poemy way.' She repeated the line stressing it rhythmically. They both spent considerable time suggesting changes and rechanges to the text. Arlene again: 'and the daylight stays for a long time. No. No...and longer does the daylight stay...' Arlene was helped by the talking and when her mother answered the phone waited for a while, then, 'Mum! Anyone! Sure is annoying. Frustrating too. Mum! I think I'll turn the tape off until my mum gets back..' She said on two other occasions that she needed help with her story. Her mother encouraged her to read her work out aloud. She kept the word processor a week longer than the other children so that she ccould continue with her writing. She finally wrote four pages of single spaced text with seven chapters called the Gandean War which was still unfinished at the time of print out. She was aware of an 'audience'. Like other children, she talked to the tape recorder. When she sneezed, 'pardon me, university'. At one session she said 'That is the end of today's lesson. Goodbye university. By the way who is listening to this? Will you phone me and tell me back?' Her mother helped with the story, with spelling and encouraged her to try out the synonym facility. She was quite precise with some words, such as cursor when she was talking to her daughter, but also talked about 'the speech things'.

Rachel (Year 5) said, not on the tape but at the end of the print outs: 'Me and my mother have enjoyed working on the lap-top computer.... and we enjoyed sharing how we live our lives...now to mum'.

'I also enjoyed working with Rachel very much and it was fun, but it was also hard work. A lot of thought went into the things we did and we enjoyed it so much that we would like to do it again if we had the opportunity.' Rachel's tape is the most 'conversational'. She and her mother had worked out beforehand the topics and they discussed the subject matter together. Perhaps that is one reason why Rachel wrote around topics and did not include poems or stories. She described school life, including a description of PE exercises, reasons for detention and her 'personal assignments'. She had a range of hobbies, in particular cooking and teaching herself French. She had learnt from a book and from her elder sister and she and two other girls at school wrote and talked to each other in French. On several occasions she wrote down the kind of 'French conversations' they had. She described how she cooked a variety of recipes, including 'the Jamaican traditional Sunday dinner, rice and peas'. Her mother listened attentively, often asking questions and commenting throughout so her personal contributions are high. 'I'm not as good as you, 'cause I can't teach myself anything.' She was extremely supportive even when commenting on her daughter's typing deficiencies. When Rachel commented that she was 'not very fast', her mother replied: ' Well, you don't have to be very fast as I can see you're only using one finger. Well, you'll have to do better as we go along'. 'I'll try and use two fingers.' In this instance, Rachel's mother clearly gave over to her all control of the word processor (see previous comment in this chapter) as she did not feel capable of giving any instructions regarding it.

All the children in this group were encouraged to read their text out loud. All the mothers and Nisar's tutor were critical about typing skills and the children's reliance on a one-finger approach. All the children took the initiative in their learning and, although they depended on others' advice and support, were clearly in charge of the word processor.

Figure 1

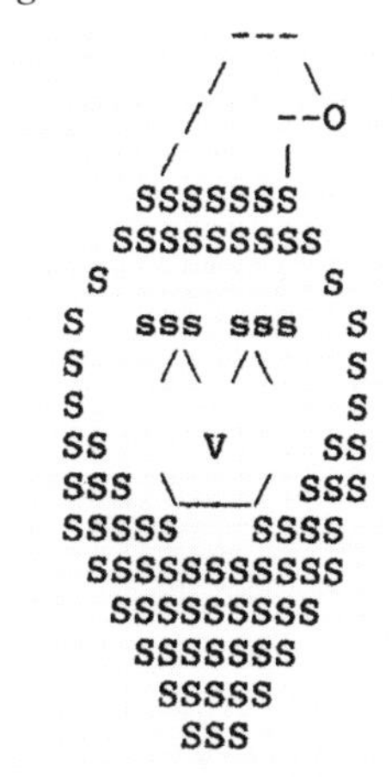

Father Christmas

Case C

Two Year 4 children, both boys, needed persuading to continue work. Both seemed very enthusiastic when observed at the print out sessions at school but they did not prioritise it in terms of their various activities. Both talked about playing computer games with their friends (which only one of the girls mentioned) and both were 'jollied along' by their mothers. In both cases, work on the word processor allowed them to communicate certain aspects of their computer interests to their mothers which they had not previously discussed. Miguel talked about his activities on a Saturday which included going into a games arcade. His mother queried: 'Did they let you in there?' in a rather surprised tone. Jacob wrote 'my favourite Beat 'em up game is Golden Axe its main characters are a dwarf with an axe, a barbarian with a big sword and an amazon with a sword. this is a violent game which my Mom would not have let me own if she knew what was the content of it. My favourite non-violent game is Deluxe paintbox 3'.

Both mothers were enthusiastic about using the word processor. When Miguel had gone round his friend's house his mum wrote 'i am sat playing on the word processor all on my own having a wonderful time enjoying every minute.' She and Miguel designed and constructed a series of 'pictures' by keying in various letters. The designs got better and better. The mother wrote: 'This time we cheated. We took a piece of graph paper and drew our picture (Figure 1), then decided which numbers, letters or signs we were going to use for each part of the picture...I think this could be a good exercise to try with the children because they have to think about their spacing, counting out the letters, the spaces and the rows. I had to give Miguel some help, but he soon picked it up.' As in other families, siblings used the word processor, and Miguel's older brother Ainsley completed a long piece of homework on his reading book, *Joby*.

The similarity of the tapes, positive and enthusiastic mothers with sons who tended to see computers as games machines, corroborated findings by Elaine Millard (1996), that some boys actually expressed a strong dislike of reading - 'they represented their reading as a hypothetical construct. That is they could read but chose not to. In other words they were alliterate rather than illiterate and their reading skills were being allowed to stagnate.' These two boys in our study did seem to choose to do other activities than writing (which is what they probably saw word processing as being). Jacob wrote at the end: 'What I think about using a lap-top It was O.K. but I didn't have enough time. And the little time I did have I spent playing with my friends.' Given that they also liked computers, but from a rather limited viewpoint, 'greater use of information technology as a way of motivating boys to develop their reading and writing' as currently advocated on their web pages by the National Literacy Trust seems not entirely to be the antidote. Their pleasure in creating 'pictures' and using paint programs could be used creatively, perhaps in creating multimedia texts.

Case D

Several children were left by themselves to work on the word processor. For instance, Goldie's mother said to her: 'carry on duck...and when you've finished most of it I'll check it all over and make sure that it's OK. Alright?' However, she must have been in earshot and even eyeshot as Goldie asked many questions to which her mother gave 'instructional' answers frequently interspersed with phatic comments. 'You can do. Good, very good. Do you want to say what today's date is?'

Usha virtually talked to herself most of the time. Her mother came occasionally to look at what she was doing. She instructed Usha about using the shift key and making a new file. On one occasion she said: 'I'm going to leave you now and I want you to type something about flowers. So whatever's in your imagination, write it down. ' Usha : 'yes'. 'And speak it out while you're typing it, so that you're recording yourself as well.' Usha talked as she was writing and said each time, 'The end. I've finished now. Thank you.' Her final comment on the tape was: 'Thank you for listening. Finished. Bye.' Her mother, as other mothers did, wrote a couple of poems herself, one on flowers. She was quite enthusiastic about the project and clearly instructed Usha each time to talk into the tape recorder. Possibly we did not communicate effectively enough the value of parent/child discussion to her and some other parents.

Conclusions

In every single case there were examples of co-operative effort, although this feature was certainly more marked in some partnerships. There was a support continuum, including parents who 'organised' the learning time and those who were positively 'negotiating' a learning relationship with their child. The parents were not necessarily at one point only of the continuum. Comments from Noreen's mother on two separate occasions indicate that she was shifting in her advisory role. 'She is very competent now at working the computer, but still resists suggestions from me about correct spelling. Perhaps that is in response to my approach.'

'One important issue ....is that I have been very aware of the stimulus for discussion that working with Noreen has created. Some of the benefit of doing this together has been the interaction between us. Another aspect is that it has made me aware of the need to develop skills in finding something positive to say about Noreen's work, before diving in with corrections.' The place on the continuum is also connected with the ability of the child to present himself/herself as a capable writer. This might imply a certain level of attainment but also indicates a differing attitude by the parent to the 'scaffolding' that is taking place.

The minority of parents who predominantly used a question/answer approach achieved similar results to the teachers in David Wood's research (1992), that is very brief often monosyllabic replies. In addition, comments on the tapes or messages via the word processor, indicated that parents and children ended up finding the sessions 'a chore'. 'Angela is not at all enthusiastic about this computer work. She gets bored very quickly with it.'

The majority of parents and children read through instructions together, tried out strategies on the computer, looked up spellings, tried to get the spell checker to work, finished off sentences for each other. Parents used phatic comments frequently, virtually always of a positive kind and often reflecting the outcome of a small negotiation which had been taking place, for instance the spelling of a word or the use of a capital letter. A number of parents did employ 'controlling moves ' at some points but seldom achieved the same effect as the teachers in Wood's research. In fact sometimes children ignored effectively or appeared to ignore their parents' questions which were often repeated. When there was some evidence that the child had in fact been listening although not making any vocal reply, for example had changed the spelling or put in a comma suggested by the adult, then the parent made some phatic comment such as 'brilliant'. The phatic comments tended to be timed to the completion of a sentence or the spelling of a difficult word, timing that would be understood by their child. The phatic comments were always quite specific, unlike those made by someone looking at a picture or after reading a complete story. Then the comments would be general. The parents in our survey were much more specific in the sense that they praised a particular item. On the other hand they also praised a number of actions in the same way, thinking of an interesting description, using an unusual or 'difficult' word, putting in a capital letter or some punctuation or managing to use the spell checker.

It is obviously extremely difficult for observers (parents in this case) to suppress comments about spelling, capital letters, paragraphs, apostrophes and other punctuation which they saw in front of them on a screen. Every onlooker in our sample did so comment. The screen, even the small one on these word processors, renders the work visible and parents know the text can be changed comparatively easily. Of course, they might have been reacting as if they were the one writing. We frequently mix levels of editing, so we might redraft and respell simultaneously. So, in an observer role they felt it incumbent on them to comment immediately they noticed an error. Additionally, many parents (with the exception of those whose children were encouraged to be independent writers via emergent writing) felt it necessary to ensure that the sentences had capital letters and full stops. Even where, with one school, they ticked a sentence saying that they had not given much help to the child, the sentences even of very young children had capital letters and full stops. Parents wanted to praise their child, as evidenced by the number of phatic comments, yet strove for a balance, as they saw it, of writing within a particular frame of presentation. More open debate amongst parents and teachers as to the benefits of word processing can encourage individual decisions as to what are the best times for 'help' and 'correction'.

None of the parents, of course, was in the normal teacher situation of having to control a class of children. Control was sited usually with the child, perhaps because they were in charge of the keyboard, which coincides with the findings in Rupert Wegerif's and Lyn Dawes' chapter that the child with the mouse often had a controlling position. The child therefore mainly controls what is actually written. For example, this pair are talking and writing about winter.

Mother: Can be wintry.

Tanya: Can be wintry...wintry...And it can be icy.

Mother: yes?

Tanya: Full stop.

Mother: Is it wintry or icy?

Tanya: Icy.

The use of the spell checker became a control issue. Parents did not find this version particularly friendly but persisted in using it although as far as we know 'no official' encouragement was given in this direction. On some tapes a high proportion of time was given over to organising its use. Children expressed annoyance at the beeping of the spell checker, often telling it to shut up. However, both Miguel and Noreen enjoyed outwitting the spell checker by finding words it apparently was unable to spell, such as 'timelines'. Both also expressed pleasure on occasions when they spelt a word correctly which they had been unsure about. The lack of a beep was a measure of their success.

Parents kept a sense of 'knowledge control' not only about spelling, but also punctuation and sometimes when talking about files or formatting text, when the knowledge could be said to be specialised. This was also true of other knowledge, for instance about family background. Their instruction was most often in transmission mode. They had 'knowledge' which they wished to transmit to their child. No parent behaved like a teacher in terms of setting out what their child was to do. The most general question, either at the beginning of the tape or when the child was beginning a new piece of writing was 'What do you want to do?' or 'What do you want to write about?' The most common instruction by far was connected with spacing and often was the single word 'space'.

The computer screen may take the place of a piece of paper in the writing process but was undeniably seen differently by parents and children. The screen objectified the writing to some extent in that both parent and child could discuss alterations in the text in front of them. The word processor, the spell checker and the tape recorder were spoken of almost as sentient beings which 'listen' and 'understand.' They were often addressed directly. One child wrote a poem to the word processor accusing it of beeping at her. This personification was detailed very effectively many years ago by Sherry Turkle in *The Second Self*. Seen from this viewpoint, the word processor was a third member of the learning partnership. So the distinction between 'content free' and other software is not watertight. Both the spell checker and the flexibility of the word processor in changing lower case to upper case and inserting textual alterations could be used by parents as part of their control strategies.

The word processor moved around, not just from school to home and back again. One mother when interviewed explained that she and her daughter chose to sit in the kitchen so that they could have a quiet place in which to work. The tapes revealed that the television was often on while they were working. Other people came by and commented both on the text and in general terms.

Adult: when did you have that, Noreen?

Noreen: it's a school thing

Adult: it belongs to the school? And they allow you to bring it home?

Mother: we're taping. She's got to do her lessons and do it in front of an adult.

Adult: Oh really. So the adults are taking over the work for the teachers!

Most parents were enthusiastic about the overflow of learning from school to home.

> I have really enjoyed working on this project with Samantha. Although I hate to admit it, bringing this project home has made me spend a lot more time with Samantha, concentrating on her reading and writing....
>
> It gave us more of a chance to share time and gave us a shared interest. Helen was a bit dubious at first even though she has used a computer since nursery, this was a bit different as there were set work to do besides some work she chose to write herself. We found it time consuming sometimes when she had other things to revise (times tables and spellings) but we used the laptop for practising both.

The examples in this chapter show small shifts towards children's independence in learning. We may not wish to proceed as fast or as far as Seymour Papert (1996) suggests: 'give everyone a computer, and then here and there more and more people will find interesting things to do with those computers and the new ideas will spontaneously grow. Maybe a horticultural model is the best one....I think this image of planting the seeds to grow everywhere is already happening...Maybe kid power will force schools to go out of existence.' Schools remain extremely useful institutions, but they become even more useful when they work with the grain and encourage parental help in the learning process. That doubtless means that more research is needed to focus on this shifting context.

I express my thanks and appreciation to all the parents, children and teachers from Southey Green, Meynell and Firs Hill Schools who took part in the project. In particular I wish to thank Ruth Crowley, Karen McSweeney and Kathleen Parker. My colleague, Jeff Wilkinson, has been of immense help both in the project itself and talking over the findings later. I thank also Francis Howlett from the NCET for his encouragement and support. The children's names in the chapter have been changed.

References

Blair, T. *Connecting the Learning Society*. DfEE: London, 1997.

Blunkett, D. *The Guardian*, July 15th. 1997.

Bruner, J. *The Relevance of Education*. Allen and Unwin: London, 1971.

Cox, M. *The effects of IT on Students' Motivation*. National Council for Education Technology:Coventry, UK, 1997.

Mercer, N. *The Guided Construction of Knowledge: Talk among Teachers and Learners*. Multilingual Matters: Clevedon, UK, 1995.

Merchant, G. & Monteith, M. 'Laptop as Messenger: an Exploration of the Role of Portables in Home-School Liaison', *Reading*, July 1997.

Millard, E. *Some Thoughts on Why Boys Don't Choose to Read in Schoo*l. Paper given at Sheffield University, May 1996.

Monteith, M. 'Combining Literacies'. In Neate, B. (ed.) *Literacy Saves Lives*. UKRA: London, 1996.

National Literacy Trust, Swire House, 59 Buckingham Gate, London SW1E 6AJ http://www.literacytrust.org.uk/

Office for Standards in Education OFSTED *A Review of Inspection Findings 1993/4*. HMSO: London, 1995.

Office for Standards in Education OFSTED *Guidance on the Inspection of Nursery and Primary Schools*. HMSO: London, 1995.

Parents' Information Network PIN, PO Box 1577, London W7 3ZT, UK

Papert, S. *Looking at Technology through School-Colored Spectacles*. Paper given at MIT Media Lab Conference on Educational Reform and the New Media, 1996.

Turkle, S. *The Second Self: Computers and the Human Spirit*. Simon and Schuster, New York 1984.

Underwood, J. & Monteith, M. *Supporting the Wider Teacher Community*. National Council for Education Technology: Coventry, UK, 1998.

Wegerif. R. *'Using Computers to Support exploratory talk in the classroom'*. Paper given at Computer Aided Learning Conference, Cambridge, 1995.

Wells, G. *The Meaning Makers: Children Learning Language and Using Language to Learn*. Heineman Educational Books: London, 1986.

Wells, G. 'The Centrality of Talk in Education'. In Norman, K. ed *Thinking Voices: The Work of the National Oracy Project*. Hodder and Stoughton: London, 1992.

Wood, D. 'Teaching Talk: How Modes of Teacher Talk Affect Pupil Participation'. In Norman, K. ed *Thinking Voices: The Work of the National Oracy Project*. Hodder and Stoughton: London, 1992.

Wray, D. & Lewis, M. 'Young children's talk during authentic inquiries'. In Gambrell, L.& Almasi, J. (eds.) *Lively Discussions,* (1996) 63-72. International Reading Association.

New Writers, New Audiences, New Responses

Chris Abbott

Introduction: the size and scope of the change

The World Wide Web has been described in many different ways by different people; in 1996, for example, by the US High Court as 'the most participatory form of mass speech yet developed' and by a 17-year-old American high school student on his Web home page as 'an outlet to describe my thoughts, ideas and feelings'. It offers a complex amalgam of publication and communication, a medium which seems to have been seductively alluring to young people from many different backgrounds. Since 1994 there has been an explosion in the publication of personal homepages: electronic documents conveying thoughts, feelings, ideas, personal history, dreams, wishes and ideological belief.

Until the advent of the WWW, publication usually meant a collaboration between a writer and a publisher, with one of the main roles of publishers being to sift and sort through the many writings submitted in order to choose those that they consider not only intrinsically of value but which might also be successful in the marketplace and thus earn income for the publishers.

Costs and income have little to do with Web publication, at least in its original form. The wherewithal to become a publisher is often packaged along with a basic Internet connection. A user subscribing to an Internet Service Provider (ISP) will probably be offered at least some Web space and possibly a very large amount of it. This Web space can be used over and over again, changed, amended, publicised and visited by thousands of other Web users.

> *Hi this is XXX*
>
> *I am Eight and a half and I live in England (Forest hill London ...)*
>
> *My intrests are Cycling/Sport/and Sleeping.*
>
> *My favourite Sport is Cricket*
>
> *My Favourite food is Sausage and my favourite drink is Coke.*

I go to XXX Prep School.. our Class is called 3c

My teacher is called Mr XXX

My favourite lessons are IT and Art, my worst lessons are Rs.

I Play the 'cello, piano and recorder.

I use computers A LOT

We have had a holiday in Wales

This has little in common with any form of publication previously available to young people or indeed most writers of any age. Before the Web, the only way in which young people could get their work published was in anthologies or other works which were usually produced by small companies and in limited print runs. The avid Web writer, on the other hand, can move from a written text to proof-writing stage and then to publication worldwide within a matter of hours or even less. It is comparatively simple for a young person to write a poem one evening, convert it to the version of text understood by the WWW and then upload it to the appropriate Web server, a process which need only take a few minutes. The page is then immediately available to anyone in the world with a web browser and Internet access. Within a matter of hours that young poet may be receiving comments and questions from readers of his work. Comparing this to the lengthy process necessary to write this book is a salutary process, and goes some way to explaining why so much of what is included in this section may be appropriate rather more for historical understanding than for illumination of current practice.

What has happened so far?

The WWW was invented in 1993, together with Hyper-Text Markup Language (HTML), the language in which Web pages have to be written. This language is not specific to any one computer platform and the Web browser software that reads the pages is usually free, or apparently so. Anyone with an Internet subscription has access to the WWW and can therefore publish on it and read what others have written. The phenomenon of the WWW Personal Homepage soon developed and has been the subject of research by the current writer among others.

By the middle of 1994, a vast self-publishing phenomenon was developing, with people, institutions, companies, schools, governments and even cats having their own presence on the Web. This was radically different from the kind of texts that had previously been e-mailed across the Internet, since Web pages are essentially seen as publications rather than communications. They deliberately have no particular reader and are designed to be read by many, semi-anonymously, although the technology allows page-owners to track down who has been reading their pages.

An early need identified by enterprising individuals was the necessity to find information on such a vast and ever-changing resource. Menu interfaces such as Yahoo and search engines like Alta Vista and Lycos were developed in order to instil some structure into an apparently chaotic resource, and they were quickly successful and became the site of much of the early advertising which began to bring income to Web page owners.

By 1996 the search engines and menus started developing into multiple versions. Yahoo offered country-specific searches with Yahoo UK or Yahoo Germany, and even a special version for children called Yahooligans. Search engines appeared which looked only in a country or region, or in a continent in the case of the European search tools which were developed in response to concerns about the US emphasis apparent on the system.

It took a long time before anyone made any money from the WWW, with rumours abounding of online sales services which cost tens of thousands to set up and then sold only a few units over many months. The development of a safe mechanism for transfer of funds electronically was identified as the major hurdle to be cleared, and much effort was put into a number of competing systems.

A further concern existed over material which might be considered offensive, inappropriate or even illegal. Various attempts were made to control this material or block access to it from particular services or sites. Much of this development was in relation to Web access for schools and young people, and I had close personal experience of this development as Educational Adviser to RM Internet for Learning. Newspaper articles about this issue and other associated publicity were influential in forming views of the resource on the part of many educators and parents. There is certainly offensive material on the Web, and many schools were helpful in identifying sites that should not be accessible, and Internet for Learning were then able to block these centrally. In many cases, acceptable use policies and research contracts were seen as the way forward by schools.

What are some schools doing to respond to this?

Early in 1995, and following the BETT exhibition in January of that year, small numbers of British schools began to publish their own pages on the WWW. During the first few months of Web presence for these schools, March to July 1995, I collected sample Web pages. These were randomly selected from the fairly small number of such pages then available. It became apparent that at this early stage, there were two main modes of use of Web home pages by these schools, as an electronic brochure, and as a virtual library of resources for students to use.

> *St Alban's is an 11-18 mixed comprehensive school maintained by Loamshire County Council. We draw our pupils from a wide and largely rural area of*

> *Loamshire immediately to the East of the City of Loambridge and from a small part of Grantshire.*
>
> *The school occupies the superb site of Loamshire Park with its Georgian mansion and extensive parklands. During the 1980s the Authority invested heavily in extending and improving the facilities, and this enabled the whole school, which until 1983 was divided between two sites, to be united on the one campus - surely one of the most beautiful in Loamshire. A magnificent Sports Centre came into use in September 1985, and this facility is shared with the community.*
>
> *There are 230 in the Sixth Form, and 1,368 pupils in all with the equivalent of 80 staff.*

The first use, as an online brochure, is by far the most popular. Just as the first CD-Roms tended to be electronic versions of existing books, so have many of the first Web home pages tended to offer another format for information previously available in a different medium.

Primary schools, as has often proved to be the case with innovative technologies, tend to be more enterprising and original in the way in which they publicise their existence. Primary School A provides factual rather than qualitative information, and then backs this up with photographs of the school pond, the school sports day sack race and other recent events. This would appear to be an excellent way of communicating with Web-aware parents about events that have happened in the school.

Like many schools in its category, Independent School A carries basic information about the school on its Home Page. In view of its particular focus, the school also provides a link to the same information in Greek.

> *We believe that Grantchester Grammar School offers a unique environment for your post-16 education.*
>
> *We believe that hard work, effective personal organisation and a happy student atmosphere are crucial keys to the considerable academic success achieved here.*
>
> *We are especially keen to offer you a degree of responsibility for your own learning and personal development, within a framework of support and guidance.*
>
> *We offer a wide range of GNVQ, 'A', 'AS' and GCSE courses, with the freedom for you to choose almost any combination of major subjects.*
>
> *We encourage students to take opportunities to develop old and new interests and skills, through taking part in, or through leadership of a wide range of extra-curricular activities.*
>
> *We work very hard indeed to help those who have special needs, whether border-line ('A' level) candidates or highly gifted students.*

One other school which has a signed message from the Headteacher is Primary School B. This small rural school has chosen to put the information about itself after the lists of links and sites which it hopes children at the school and elsewhere will explore. More Web-aware than many of the other pages, and constructed with the aid of an outside consultant with particular expertise, this page is representative of a growing number of pages fitting into the second category, those which function as an online research centre or virtual library.

RM plc, who provide Internet access through Internet for Learning, have made server space available to schools, and it is often the most IT-aware and experienced schools, such as Independent School B, that develop this area first. Unlike all the other school pages viewed during early 1995, this school chose to include student pages where young people could present their own views and responses. This site seemed likely to be one to watch closely since it appeared to be innovative in its use of the technology, although sadly when revisited many months later no changes were apparent and the early enthusiasm had not been maintained.

Failure to change and update Web pages is an indication of a lack of understanding of the potential of the medium, and is likely to lead to readers failing to return to the site. Those people who do update their pages regularly have taken to including a piece of coding which displays a form for regular readers to fill in; they will then be automatically e-mailed every time the page is updated. There is no equivalent of this in any other medium. Acceptance of the value and necessity of changing content is part of the process of becoming a Web writer; just as the first CD-Roms ultimately failed because they were simply books viewed electronically, so the brochure-type pages signally fail to understand the medium.

Development in schools may have been rapid, but the pace at which young people have seized on the opportunities for Web publication has been remarkable. From the earliest beginnings of the medium, young people have chosen to publish their thoughts, feelings and interests. They use their Web pages to link to their favourite television programmes, to write about the music they like and to invite others with similar interests to contact them. Many young people publish poetry and fiction in this way; others feel driven to share their life stories with the wider online world.

In early 1996, I contacted over 70 young people aged from 12 to 25 years who had homepages. Finding these young people was not easy; although lists of homepages existed, it was not always easy to deduce from the pages themselves the ages of the authors. However, 47 responses were received from young people in the target age-group, which is an astonishingly high response rate for a questionnaire, 67 per cent. It was, of course, much easier to reply to this electronic questionnaire than it would have been to overcome the inertia which might otherwise have stopped them from writing on paper, putting the results in an envelope and posting it. E-mail questionnaires involve only a few on-screen ticks and then the click of a mouse to send the completed sheet on its way.

Half the young people involved were living in the USA or Canada; the rest were in the UK, the Czech Republic, Finland and the Netherlands. All the Web pages were in English, and this has caused some concern on the part of observers in some countries who are worried about the future position of national languages which may be spoken worldwide by comparatively small numbers of people. Almost all of the young people in this random sample were male, although the 11 per cent of females indicates a state of play in those early days rather than the current likely distribution. It is widely believed that a greater proportion of women are involved with the Internet and with Web publication than has often been the case with previous information technologies, but this is an area where research is needed to establish the parameters of the change.

This might mean that minority language speakers who are geographically separated could use the Web to keep in touch with each other and to form and maintain virtual community groups. By late 1996, interesting developments in this area could be seen in the availability through services such as Geocities and Webring of virtual communities linked by interest, background or even sexual orientation, to which homepage owners can declare allegiance and be listed together.

Almost three-quarters of the young people claimed to update their homepages regularly, and most of them had not had a page in existence for more than six months at this time. The younger members of the group tended to have simple pages with lists of interests such as pets, football and family. Young people in the post-16 age-group, particularly those in higher education and therefore with easy access to the Web without cost, tended to cover more discursive topics on their homepages and to use the opportunity to discuss topics on which they have strong feelings or about which they are concerned.

These topics vary widely, from a plea for the missing Transformer toy needed to complete a collection, to advice for other young people who are gay based on the page owner's personal experience. It is one of the most fascinating aspects of Web homepages that both the examples just quoted come from the same Web page of a 15-year-old youth in the USA. It is not, of course, surprising in itself that someone who is gay is also a keen collector of Transformers; what seems more surprising is that this young person's literary practices include placing his Transformer-collecting alongside his sexuality as facets of himself which he wishes to publish on the Web. Many young people who are motivated to argue for change in society have also become active on the Web, and there are many pages arguing passionately for saving the whales or the rain forest, or dealing with other environmental or social issues.

Perhaps the strongest link between this diverse group of young people was in their motivation for maintaining a page in the first place, for 78 per cent of this group the principal motivation was communication with others. Very few of the young people were aiming at readers of a similar age, and it is striking that age, race and class seem to have little importance on the Web, at least in its mostly text-based early form.

Race is, of course, invisible, especially where pseudonyms are used for names. It would be interesting to know the extent to which the names chosen are or are not indicative of ethnic origin. I have had online discussions with several young people who have told me that they chat on the Internet with a much more racially diverse group, they suspect, than would be the case in their high school or neighbourhood. However, these were all young people in America and it may be that this is a phenomenon which is much more noticeable in that setting. Within the UK, accent has always been a predominant class marker, but of course it is not available when text-chatting. The coming upsurge in the use of Internet telephony is likely to change all this however, and many young people are also beginning to add sound files to their pages so that their readers can hear their voices. Class is often hidden from view, at least in the UK where it is closely associated with accent, although hobbies, interests and even styles of page can reveal much about aspirations and identifications.

It is certainly true that most Web page owners are of university age or younger, although this is self-selecting in the case of my own research as I have been working only with my target age-group of 12 to 25 years. The Web does, however, seem to have an enduring capacity to make age difference less important. When I need assistance with my own very basic Web pages, I often go with a genuine need to be taught, to a young person of 15 or 16 years. Like many other individuals and companies, I have also employed young people as page coders and designers, and have given them considerably high status within projects. When teaching an opportunity class for 7-10-year-old children in south London recently, I asked two sixth-formers to work with the primary children to create Web pages. This they did very happily, and it was startling to see a 17-year-old young man quite happily work to the direction of a 10-year-old girl whose page he was helping to design.

The 37 per cent of the group who were aiming at a readership of approximately the same age were often the youngest and those who had included an abundance of photographs and other personal detail. Many of these young people exhibited a high level of self-esteem and confidence, perhaps a necessary corollary for someone prepared to publish what is essentially a personal manifesto to the world.

As new technologies became available, it was often on the homepages created by young people that these were first seen, or seen in innovative or inventive uses. Perhaps the first of these developments was the addition of moving rather than just textual images. Animations gradually gave way to video clips, and pages soon became moving, talking and musical multimedia sites.

The addition of a link to a Guest Book added considerably to the interactivity possible on a site. This software, often resident on a distant server on which the subject registers a presence, enables visitors to the homepage to register a response, answer a question or raise a further issue. Chat software, where connections are fast enough, enables visitors to talk in real time with the owner of the page and other guests looking at it at the same time.

Audio became easily available on Web pages during 1996. Where previously only downloadable sound clips could be found, gradually the abilities of software such as RealAudio enabled streamed sound to be produced, offering almost real-time transmission of distant radio stations or low-cost telephony. Internet Phone and other telephony services rapidly became a major challenge to traditional telecommunications providers. Limitations of sound quality and functionality in the early days were of little importance alongside the considerable attraction of extremely low-cost international calls.

CUSeeMe offered much the same facility for low-cost video conferencing. The picture quality was poor, the sound sometimes non-existent; but many young Internet users were more used to conversing in typed text than in audio in any case. Consequently, they were quick to add cameras costing less than £100 to their computers and video links to their Web pages.

What should all schools be doing in future?

By 1997, the WWW offered schools a medium of sufficient maturity to provide a publishing base for truly multimedia texts at surprisingly low costs. Although a few schools began to develop this resource and explore its potential, for many this was constrained by lack of understanding of what is possible or knowledge of the technology itself.

Since very few schools understand the essentially cross-platform nature of the WWW it is hardly surprising that they have been slow to realise that the development of HTML has offered them the ability to publish work in a way which can be read on a range of computers. A small poetry anthology, for example, could be encoded with HTML, images drawn by pupils could be added, and the resulting files compressed to fit on one floppy disk. All pupils with a computer at home, whether it be a Mac, a PC, Acorn or even an Atari, could then have a copy to take home and view. Since Web browsers are mostly free, and no Internet connection is needed to view files off-line in this way, the school is able to publish work in this way at minimal cost, at least to those homes owning a computer.

We are at, of course, a transitory stage on the way to widespread Internet access in the home, a development likely in the next few years. Schools need to plan for this by beginning to archive material produced by pupils so that the multimedia school resource sites of a few years from now will have a rich collection to draw upon. Artwork, music, texts and even videos of performance can all be stored digitally, as can information relating to the history and development of the school.

It seems likely that virtually all schools will have their own homepages, although these are likely to develop quickly into full-blown Web sites with many different uses. There will probably continue to be a place for the online brochure, but it is likely to be supplemented in future with video clips, sound files from the Chair of Governors, and e-mail links to various departments. The most interesting developments, however, are likely to come in the area of schools' use of Web sites as resource centres.

Many schools have begun to consider in recent years the role of the library, the resource centre and perhaps the reprographics department. They have seen the gradual convergence of these facilities, with ICT providing the link which unites them all. Prolonged attention to this area has often led to plans to develop multi-purpose resource areas with digital information sources in the form of CD-Rom and online services. The natural development of this trend is the notion of a school Intranet (Figure 1).

Figure 1

Check out our SELECTED LINKS page. This is a short list which will be changed and added to regularly with places we've explored and found to be useful and fun to visit for both children and teachers.

We use the Internet to help us do research on the TOPICS we do each term.

Sometimes we do interesting things and we like to write about our experiences and share them with everyone.

We have some pages dedicated to traditional festivities.

Where the Internet is a catch-all term to cover the vast and anarchic conglomeration of computers linked by a diverse collection of telecommunications systems, Intranet was a term coined in the mid 1990s to describe an on-screen information and communication system which operates within one institution and is closed to the outside world. Although the business sector developed this notion first, schools have begun to see its potential and some have developed interesting uses for it.

Personal homepages are likely to be a continuing factor, with ever higher percentages of pupils having these and able to use them to link with each other and with the wider community. The link between school work and homepages was first apparent in the activities of those students who chose to make available to others the essays for which they gained good marks. These essay banks, as they have become known, are a fertile source for some students whilst forming a considerable area of concern for the teacher anxious to recognise plagiarism.

righto... diz is some of my gcse coursework for those of you who are increadiby lazy (and very stupid) - if you're half decent at school then you'd be better to do the work yourself coz you'd get a better grade than i did... but on the other hand... if you're thick or in 3rd set or below and think you'll get D's... either copy diz down or change it for yourself. happy cheating...

Conclusion, where is this all leading us?

The WWW has the potential to change the way that society views publication not only of the written word but of all forms of media. It can put in the hands of the consumer the publishing tools that have previously been controlled by only a select few. Students of the next few years will not only be able to choose what to write about, but who the audience should be, so it will be even more important that we ensure that young people understand the ways in which texts need to be modified for different audiences and delivered to them in different ways.

Media education, often seen in a consumer model, will become an active research area, with students able to monitor the effects of media texts they have published, rather than viewing the texts of others with themselves as mere consumers. Thanks to the WWW, students of the early 21st century will be active media producers as well as enlightened consumers of the digital world.

The Value of Passive Software in Young Children's Collaborative Work

Helen Finlayson and Deirdre Cook

Introduction

This chapter centres round our research with primary school children working on the computer and on similar tasks away from the computer, to investigate the contribution to the learning situation made by the computer itself. We used a passive computer application which could closely replicate a 'table top' activity involving the movement of attribute blocks.

This work is set in the context of the developing role of computers in education and of our ideas on active and passive software which we have found useful as a way of looking at educational applications. Our research suggests considerable value in the use of passive software for collaborative problem solving and we go on to discuss the implications this has for the teacher.

Group work

Children are often seen to be working in twos and threes at the computer in the classroom. This came about originally because there were so few computers in a school that sharing the task amongst a small group enabled more children to use the computer in a limited amount of time. For teachers who used group work within their normal classroom activities, such a computer approach could be accommodated easily. However for others who preferred a whole class teaching context such integration was more problematic.

The use of computers in schools

Initially drill and practice programs for basic word recognition, spelling and arithmetic were the most commonly available types of program to be used in schools and dominated classroom work for some time. Some early pioneers however, began using the simple programming language *BASIC* with children, because of the flexibility and creativity which could be developed. As soon as it became available, *LOGO* programming then became more common for its intrinsic mathematical content. Other types of educational software were developed commercially, notably adventure games (from the home computer games market) and framework programs (see below). Adventure games were quickly recognised by teachers as highly motivating and as suitable vehicles for problem solving work, involving as they do, many opportunities for exploring factual aspects of the curriculum area at the same time. Framework programs were content-free activities or games. The teacher could easily add appropriate content to suit the particular pupils and curriculum area required. Such programs came about partly as a 'watering down' of industrial software, and partly a development from the field of special needs teaching which required software to be amended for pupil accessibility. *TRAY*[1] , a text revelatory program is a classic in this genre. More recently graphics packages have come into their own as the advent of larger memory computers has given rise to greater possibilities with colour. Sound too, is now presenting exciting new opportunities, not just for the computer to produce, but for the users to incorporate into their own work.

Clearly such an expanding range of computer applications cannot be considered as one single activity. What the computer 'does' and how children respond to it are very closely related to the software which it is running, how powerful it is and the extent to which the control is retained by the programmer or given to the user. Chandler (1983), Papert (1980) and others advocated that programs should be used which gave the locus of control to the user so that they were empowered to take responsibility for their own learning (open-ended programs).

In the past there was a tendency to prefer the most powerful applications which presented activities which could not be done without the computer. Now there are so many powerful multimedia applications incorporating sounds, video and photographs and using sophisticated search mechanisms, simple choices no longer seem available. We feel there is a need to rethink the use of software in education and re-evaluate it for its potential pedagogic contribution rather than its degree of sophistication.

Active and passive software

Many of the early published studies assessing the value of the computer to pupils' learning used software such as *LOGO* programming, spreadsheets or adventure games. These are all open-ended, to some extent, in that the users can select from a wide choice of actions and communicate their preferences to the computer, each action producing a different computer response. For instance within an adventure game involving exploring a building each move to left, right, up or down could lead to a different room, possibly presenting different problems to be solved. The computer software actively responds to the users' commands in a way which they cannot anticipate. In *LOGO* programming, users are often trying to produce a particular effect and hypothesising that their actions will do just this. The software then interprets their commands producing an effect which may or may not confirm their hypothesis.

Such powerful uses of computers may not be easily replicated by other means. These then are active programs, that is programs in which the users gain more than they put in.

Other software, such as simple word processors, draw programs or information handling packages (not including graphing of data) play a more passive role. Here 'passive' is used to describe the level of computer contribution, limited to the organisation and display of the information fed into them by the user. In passive usage the computer liberates the user by making the arrangement tasks much easier and quicker to perform without adding any other contribution or dimension to the task other than a constraining framework.

The active/passive dimension is independent of the open-ended/closed one. In the above examples *LOGO* programming is more open ended than adventure games, whilst both are active and powerful. A drill and practise program which checks or provides 'the answer' could be considered to be both active and closed, whereas a wordprocessor may be relatively passive but open-ended. This is shown below in Figure 1.

Figure 1 Active/Passive and Open/Closed Software

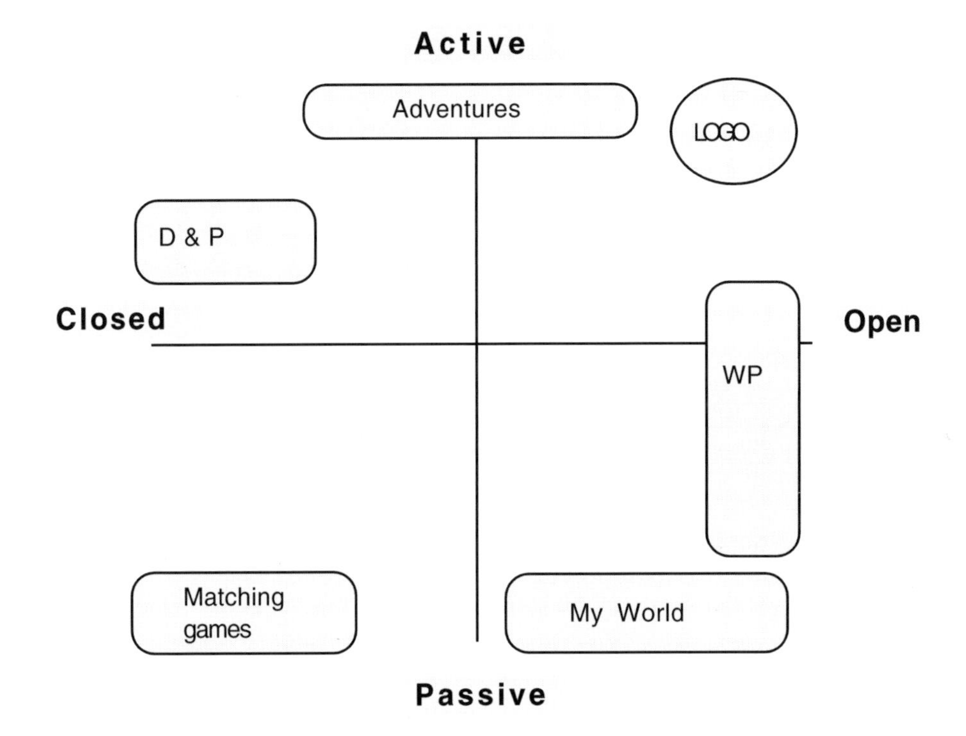

Although our early ideas suggested that the computer should not be used for anything which could be done without it, many of the current popular and apparently successful programs used in primary classes, such as simple word processors and the *My World*[2] programs seem to us to be relatively passive and do not supply answers, corrections or evaluative feedback. They only interpret or display the information given to them.

Anecdotal reports suggest that these activities with passive software are educationally worthwhile, that in fact the educational value of software is not directly related to the power of the application used. Although the activity could be completed without the computer, it may be that its intrinsic nature is actually changed when the computer is used. To make an analogy, the first bridge to be made from iron, at Iron Bridge, was made by bending long strips of it into arches, as though it were wood, because the builders only knew how to build out of wood or stone. The unique properties which iron brought to bridge building were not initially considered. So it is with computers. For example, word processors were often considered too difficult for children to use, because of the need for appropriate key use. In fact these machines were used for copy typing or making a fair copy rather than as tools for creating, drafting and editing. Likewise early graphics packages were compared to 'real paint' and criticised for their lack of subtlety and shading, rather than seen as a means to play with and manipulate a wide range of colour combinations and effects in a short time. It could therefore be asserted that passive software reflects the users' ideas back to them for their consideration, and in such a way can aid the creative thinking process.

Having explored the evolution of software to emphasise the advantages of passive software we now want to consider what active software can contribute in educational contexts. Because a powerful active application may do so much for the user it may not always be appropriate to use it. That is to say we feel it to be very important that learners are doing the thinking and decision making, not leaving it all to the computer and accepting its answers. It is important to keep initiative and control with the users, not with the machine or the software programmer. This could be viewed as a triangle of control between the users, the software developer and the machine.

With very powerful programs it is quite possible for the users to leave the control with the software developer and machine, and just do what comes easiest. The software becomes 'the authority' seeking to entertain and amuse rather than stimulating thinking and questioning which can be hard!

Figure 2 Triangle of Control

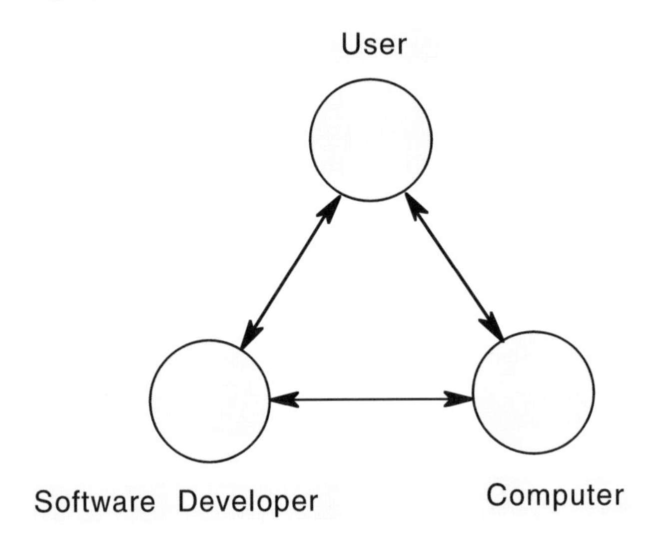

Our study

In this study we wanted to investigate the contribution which the computer itself made to the interactive behaviour of the children, to get a clearer idea of the part played by passive software in real classrooms. The application has been deliberately chosen to be clearly passive and lacks any obvious creative interpretation as in some of the examples given above, e.g. art packages or word processors.

We chose a set of tasks using attribute blocks which could be replicated on computer using the *My World* program. The computer was used in a passive role, giving no feedback other than the visual display on the screen, so that the contribution of the computer, rather than the value of the software, could be more readily identified. We were trying to find out if the computer did change significantly the nature of the task. We attempted to do this by comparing children's responses when offered two tasks as similarly constructed as possible, but with one presentation on screen and the other off screen.

The attribute blocks we used are mathematical apparatus frequently found in primary classrooms. The particular ones we used consisted of five different shapes, triangle, square, rectangle, hexagon and circle, in three different colours, two sizes and two thicknesses. We were unable to represent the thickness dimension satisfactorily on the two dimensional computer screen, so we used only the thin blocks.

My World is a graphics framework program which allows objects to be arranged on the screen through the click of the mouse. An object can be picture or text and can be picked up and put down with single clicks. Some can also be designated copyable objects, in which case each click on the object produces another copy of it which can be positioned anywhere on the screen or thrown away in the bin. In addition text can also be created by clicking on a notepad symbol and then typing in the required letters. Once created it behaves like any other object on the screen.

There are a large number of pre-prepared *My World* screens covering a wide range of the primary curriculum, including simple picture word matching, tessellations, geography weather maps, foreign language screens etc. In addition teachers can create their own screens for any particular topic, without requiring any specialist programming knowledge. The program represents to us an example of good passive software. There are no right answers, but children can arrange and print out their screens and discuss them later with their teacher. This program was chosen as suitable for representing the attribute blocks on screen and each of the 30 blocks was produced as a movable object used with a variety of backgrounds. These included a plain screen with a grey rectangle to hide blocks under for Kim's game (a memory game where one or more items are removed), a single oval for creating one set or playing dominoes against, two sets and two overlapping sets.

Figure 3 My World screen with some blocks

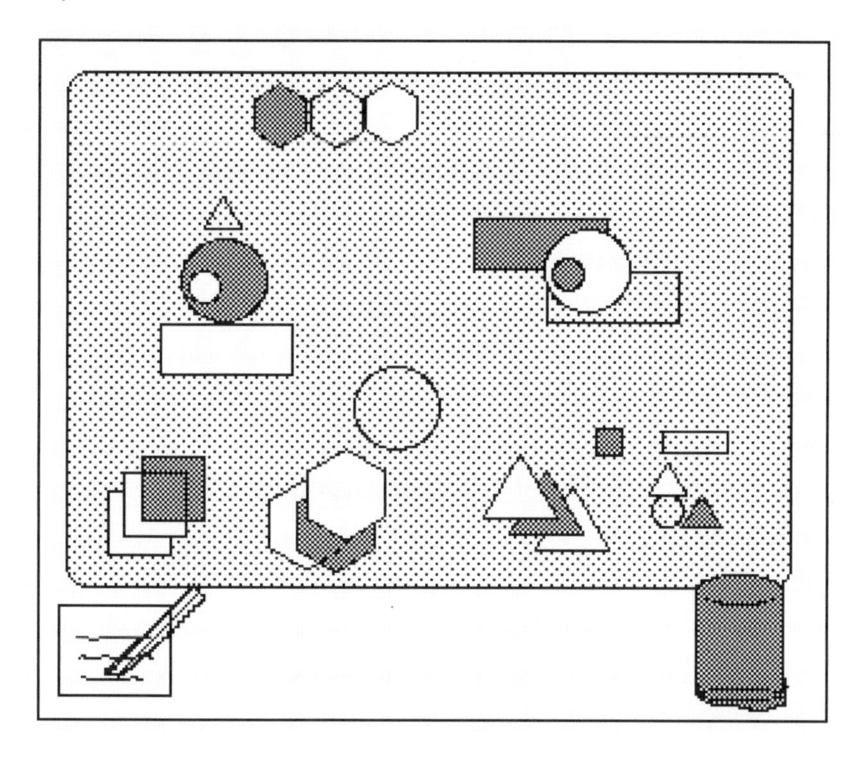

The attribute block task was chosen because we thought it was suitable for a wide age range of children, from nursery through the primary years. It was also well understood by the teachers with whom we were working. The tasks were open-ended in so far as children could demonstrate more sophisticated techniques for classifying and generalising their results, without any obvious right or wrong answers, whilst at the same time no child would experience failure or be perceived by their partner as failing.

In the initial phase of the study we worked with girls from school A, drawn from the reception class (5-year-olds), Year 2 (7-year-olds) and Year 4 (9-year-olds). This was carried out at the beginning of the summer term when the youngest children had been in school only a few weeks. Gender has been recognised as a significant dimension in both computer work and language development (Hoyles 1988, Underwood 1994), so the decision was made to focus initially on girls. They were paired by their class teacher, for the study, on friendship and perceived match of mathematical ability.

The second phase of the study involved 16 boys and girls from school B working in single-sex pairs. This group consisted only of 7-year-olds but followed the same procedures as before.

Figure 4 Children on the project

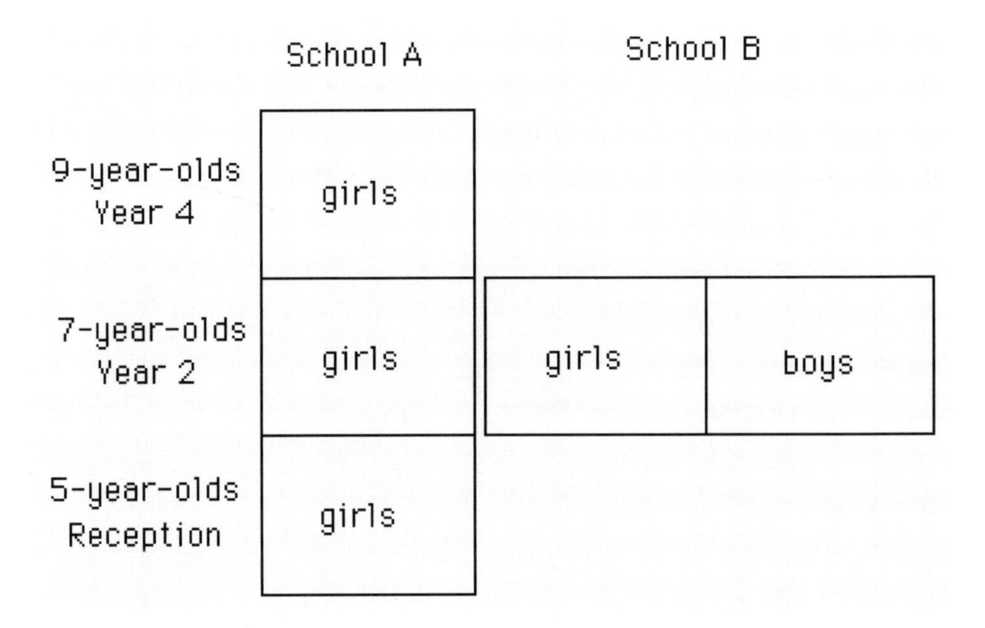

Each pair experienced working in alternate sessions, on and off the computer at the attribute block tasks, with some pairs working initially on the computer and an equal number working initially at the table. All the tasks were carried out in the area designated for practical work within a general open plan situation, where the computer was normally used. The sessions, either on or off the computer, lasted for approximately 30 minutes, determined by interest and classroom constraints. The children were tape recorded throughout using a pressure zone microphone and observational field notes of their actions were made. Throughout the study the adult role involved facilitating as well as observing, following a similar progressive task sequence with each pair.

On the table the blocks were set out in front of the children, within their reach, and moved by hand. On the computer screen the blocks were laid out in a similar manner, but moved by putting the mouse pointer on the chosen block and clicking the left button once to pick it up. The block was then moved with the mouse and placed in a particular position by a second click of the left button. An area of the screen in a different colour was used to collect the chosen blocks, and this was represented by a large sheet of coloured paper on the table.

Each pair of children had up to six sessions. At first they were introduced to the blocks, either on the computer or at the table, and encouraged to select and describe the attributes of their specific block. This was used to establish a commonality of vocabulary, and the ability to recognise all the attributes.

They then played a version of Kim's game introduced by the observer. The chosen block was physically removed from the table; on the computer screen it was moved and covered by a large grey square.

Subsequently the children were introduced to a sorting game where the choice initiative lay with them. They worked in a cooperative mode in which they created sets between them, taking it in turns to add blocks to the set. In the later stages the 7-year-old children working on the computer composed a statement describing their set, and were able to print this out. The children continued with more formal and progressively demanding sorting tasks, involving two sets, intersecting sets and subsets.

Results

As one might anticipate there were some predictable differences between the age groups. The 9-year-old children found the early tasks quite easy to do so moved through the different activities quite rapidly, and consequently had fewer sessions. When we examined the talk transcripts it was clear that these children were quite familiar with the style of classroom discourse outlined by Edwards and Mercer (1987). That is to say they tended to behave according to their understanding of the ground rules for interaction with a 'teacher'. They were very much task-directed and tried to do only what had been suggested, to the point of copying both the language and actions of the adult. For example if the adult started off a set demonstration using yellow as a defining attribute, the 9-year-olds also constructed sets with colour as the defining attribute.

The 5-year-olds, on the other hand, were newly into school and had less fixed ideas of what was required. They enjoyed the screen based activities as a game and identified for themselves a side of the screen, as well as being very strict about turn taking with the mouse. The talk transcripts here show a radically different pattern.

The range of tasks used seemed more appropriate for the 7-year-olds in both settings. They found more of the tasks challenging than the 9-year-olds and had had more experience of using the materials and the computer than the younger children. Their talk seemed to indicate a need to articulate their thinking and actions to their partners. For this reason we chose to repeat the work in a second school, for Phase 2, with both boys and girls, using only year 2 children, to check the comparability of the behaviours, as well as exploring the behaviour of boy pairs under similar situations.

Effects of the computer

Despite these differences in performance at the various ages, some common behavioural patterns were noted when comparing the children's work at the table or on the floor with actual blocks, and their activities in the computer-based task. These findings were common for all the groups of children regardless of age or gender. When working on the computer:

- the sessions lasted longer
- the children were more involved in their partner's work
- they were less open to general distractions
- they developed the tasks further or showed greater persistence
- they showed greater enthusiasm for the task.

Sessions on the computer in all comparable cases lasted longer than those working with blocks, and often had to be brought to a close by the participating adult at break or lunchtime, against the children's wishes. In some cases the interest in the computer work was sustained for more than twice as long, as evidenced by the taped data.

Moving blocks on screen took longer than on the table as it required a greater degree of manipulation, but the children also insisted on continuing the task until they felt that it had been completed to their satisfaction, for instance clearing all blocks back to their original positions, or lining them up with geometric precision, a small shape inside a large one. Our observational data suggested to us that the notion of 'closure' may be of considerable significance in computer tasks. Children were not prepared to leave the screen in an unfinished state, whereas they would happily leave the blocks on the table in whatever position they had taken up.

The children always took turns to use the mouse, often instigating this turn taking themselves without any adult suggestion. However when one partner was using the mouse, the other would be following all the movements on the screen, pointing, making suggestions and asking questions about it. Such screen watching was also shown by the rapid rate with which they were able to spot small changes, for instance when playing Kim's game.

This concentrated interaction was not observed when they were working with actual physical blocks, as they tended to divide the blocks between them and work less in a turn-taking way. Off screen presentation also generated less enthusiasm and persistence, and the children seemed more open to outside distractions, though the girls in particular, in both schools, were very rarely off task. In both presentations there was a tendency for children to create pictures and patterns with the blocks, demonstrated most often on the computer.

With reference to the composition of text, this was thought to be too demanding for the youngest pupils and not sufficiently challenging for the 9-year-olds. All the 7-year-old children were invited to make a statement about their sets on the screen and print out both the set and its corresponding text. In order to create an agreed form of words the children had to not only articulate their ideas and achieve a shared understanding but also arrive at a consensual statement.

Figure 5

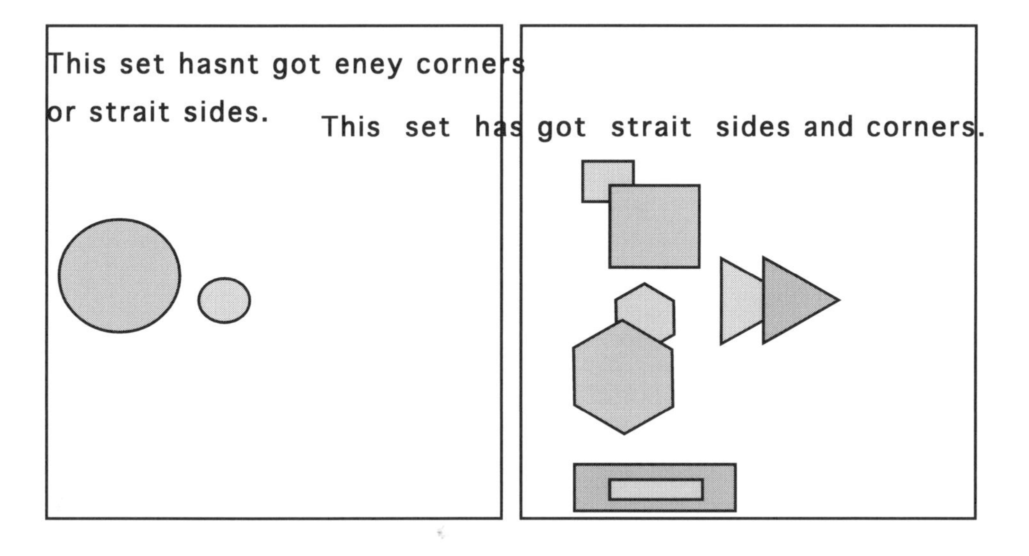

The boys, on the whole, behaved in a similar way to the girls from both their own and the other school. However, when they attempted to compose a description of their set on the screen some interesting differences emerged. Whilst dealing with block manipulation using the mouse the boys were careful to observe the conventions of turn-taking, but the response to the keyboard was different, especially with more evenly matched pairs. Here within several pairs of boys a more competitive behaviour emerged resulting in the sentence writer trying to protect the keyboard from the predatory behaviour of his partner. That is to say the waiting partner 'helpfully' pressed CAPS LOCK, space bar, RETURN or DELETE keys whilst the other was trying to write. The easily identified 'guarding' behaviour by the writer rather than openly protesting and alerting the teacher, suggested that this was 'normal' behaviour between the boys. They also tended to stand whilst using the keyboard and position themselves at either end of it. None of the girls behaved in this competitive way. Their behaviour in both schools was unfailingly courteous and helpful. The boys also betrayed some of their competitiveness through mildly negative comments during the set building activities 'I'm going to do a better one than you!'

Effects of the computer on the children's behaviour

Overall the findings which we have listed above are in line with current research in the area (Scrimshaw, 1993; Underwood and Underwood, 1990) in so far as it is possible to compare studies. Perhaps the most significant finding is that the computer does make a difference. That is to say because the tasks were very similar in their cognitive demand and the adult presence and support was consistent in both presentations and the pairings of the pupils remained the same, only the medium of the task varied. In fact one could argue that the medium was radically different in each case. We want to reflect upon these differences and consider why each effect that we noted was produced.

The two main differences which manifested themselves in the computer based tasks relate to the shared visual field provided by the computer screen, and the single input channel offered by the mouse. As we described previously when the children were working on a table they divided the blocks between them, and thus concentrated primarily on their own area of the workspace. Such an approach was impossible with the computer. The children positioned themselves so that the screen was comfortably visible and accessible to each of them, usually sitting next to each other. They concentrated on the screen and its display. The transcripts from each group show a considerable number of implicit references to the screen with comments such as:

'that looks like a .....' 'put that there'

In this particular learning context the computer acted almost as a mirror reflecting their actions back to the children, and it seems reasonable to assume that this intensity of focus played some part in maintaining the levels of concentration of the children, and sustaining them in the task.

As well as the effect created by the common visual display screen, the single input channel provided by the mouse also coerced certain behaviour. That is to say that moving the blocks on screen could only be accomplished by whoever had control of the mouse. The children passed this back and forward between them or adjusted their body position so that each participant could manipulate the mouse in the space between them (left hand / right shifts). This in effect 'forced' them to collaborate in the completion of the task in a physical sense. The intensity of their concentration and focus on the screen however also produced collaboration in the intellectual sense, a perception reflected in their comments on the talk transcripts. Interestingly too the example of more competitive behaviour in the work with the boys occurred when using the keyboard which offers several possible input channels and not in the more restrictive practices necessitated by mouse manoeuvres.

Though the tasks were very similar, and the adult participation role remained the same, in practice the experiences offered to the children by the two learning contexts were quite different. In one they took turns in playing games with blocks, in the other they worked collaboratively with a common output and negotiated shared actions.

Extending findings to passive software in general

The two elements of screen focus and single input are quite clearly important in other passive software applications when used with small groups of children. The constraints provided by the allowed actions within the software demand negotiated understandings from the children which may prompt higher levels of thinking. By giving responsibility to the users for what they produce, with open-ended software, but constraining the way they may achieve it, they are forced into thinking their actions through.

Griffin et al. (1993) suggest that the computer can act as a medium or a dialogic partner. In the latter case it provides another voice in the conversation rather in the manner described earlier in the section on active software. Such a voice can be directive and didactic and with culturally loaded content, for example with gendered imperatives (ref to pirates and honeybears). In comparison passive software can be seen more clearly as a medium, with a minimal dialogic role only.

We find the categories of active and passive software useful in thinking about educational applications. They need to be developed further and tested out more thoroughly with more research.

Our main conclusion from this research is that passive computer applications can encourage productive cooperative working, generating more enthusiasm and persistence than similar tasks carried out off the computer. This is not to say that active applications cannot similarly develop cooperative working, as has been known for some time (Hoyles, Healy and Sutherland, 1991; Underwood and Underwood, 1990), but there are two points to be made here:

1. It is not necessary to use active software to develop enthusiasm, persistence, and productive collaborative work.
2. Active software does not necessarily produce effective learning situations, since the software can do too much, taking the initiative from the users and allowing them to produce meaningless impressive effects, (Govier 1996).

The voice of the active software may be too strong to resist, so that both the teacher and the children abdicate the responsibility for the learning outcomes to the software developer. Whereas passive software leaves the responsibility for the actions with the children and the teacher.

The teacher has an important role to play in the use of passive software in essentially setting up appropriate tasks for the children and matching the level of challenge to their abilities. In our study the 9-year-old children did very little more on the computer than off it, because the tasks were really too easy to produce a challenge for them. The level of discussion between them was reduced because there was no need to articulate their understanding, only when one partner made a logical error would the other point out the inconsistency and, if necessary, explain the underlying rule to her partner.

Implications for practice and further research

We feel that the time spent doing a detailed analysis of the software and the potential it offers for children's learning is always going to be worth while, especially when the activity is going to be carried out without an adult being present. The dimensions outlined in Figure 1 need to be considered open/closed and active/passive.

We feel that it is important not to dismiss the less active software as our work suggests that it has a valuable contribution to make to learning, especially in terms of focused activity, cooperation, and articulation of ideas.

Clear definition of the learning outcomes to be achieved as a result of the computer task are essential and the teacher needs to be very clear precisely why the computer is being used. Experienced practitioners are well aware of the need to prioritise certain aspects of learning on specific occasions for particular children. This is equally true for computer tasks, for example as in this study the passive software had plenty to offer in terms of cooperation, and the articulation and formation of ideas about sets, but perhaps not much in the way of new knowledge or achieving a prescribed output. On another occasion these latter learning outcomes might take priority over cooperation, for example. The teacher's analysis and decisions are the directing force.

It is as important for pupils to understand clearly what the teacher requires them to achieve, in a computer task, as it is in any other aspect of classroom learning. In the complex pressurised environment of the classroom this is something that can be easily overlooked, especially in relation to computer work where there seems to be an expectation that the computer will make the learning outcomes clear, as it tends to do for the procedural instructions. If cooperation is the chief desired outcome then the children need to know that and have a part in assessing their own efforts in this respect.

Such reflective analysis would ensure that the monitoring of computer work by the teacher would not require their continuous presence. With clear identification and communication of learning outcomes considerable responsibility has been given to the pupils. Intermittent teacher support should be sufficient to ensure pupils working towards the identified goals unproblematically. Children should also be involved in the record keeping and assessment processes. For instance if 'talk' were an important learning outcome then children could be involved in making judgments about their own and each other's contributions in terms of whether they both had things to say about the task, who had had a good idea, whether they had listened politely to their partner's idea, and so on.

The grouping of pupils to work on the computer is still a hotly debated topic discussed by Jean Underwood in this book. The majority of studies show that girls prefer a collaborative approach to learning and thrive in small group situations with other girls, whereas boys can be successful in either competitive or collaborative situations. However girls seem to do less well and are less happy when paired with boys.

In our study we deliberately paired children by gender and for close similarity in ability. Both schools, one suburban and one rural, were broadly comparable in terms of the socioeconomic background of the pupils, so many possible variables were not considered. Our findings were in line with the general conclusions of other studies on both the cooperative nature of the girls and the changes of behaviour brought about through the use of the computer. Our study also went a little way towards investigating the mechanisms through which such collaborative changes come about.

Notes

[1] TRAY, originally entitled Developing Tray, and later Infant Tray will take any passage of text as the subject matter and initially reveal only punctuation with selected letters. Users have to use the contextual and syntactic clues to guess the missing letters and make sense of the passage

[2] *My World* is a graphics program in which drawn or text objects may be picked up and moved by simple mouse clicks. The rearranged screen may be saved or printed out.

References

Chandler, D. *Young Learners and the Microcomputer*. Open University Press: Milton Keynes, UK, 1984.

Crook, C. *Computers and the collaborative experience of learning*. Routledge: London, 1994.

Edwards, D. & Mercer, N. *Common Knowledge*. Methuen: London, 1987.

Govier, H. Paper presented at WCCE Birmingham, 1996.

Griffin, P., Belyaeva, A., Soldatova, G. & the Velikhov-Hamburg Collective 'Creating and reconstructing contexts for educational interactions, including a computer program'. In Forman, E.A. Minick N. and Stone C.A. (eds) *Contexts for Learning*. Oxford University Press, 1993.

Hoyles, C. Girls and Computers. *Bedford Way Papers* **34**. (1988) Institute of Education, University of London.

Hoyles, C., Healy, L. & Sutherland, R. Patterns of discussion between pupil pairs in computer and non-computer environments. *Journal of Computer Assisted Learning*. **7** (1991): 210-226

Papert, S. *Mindstorms*. Harvester Press: Brighton UK, 1980.

Scrimshaw, P. (ed) *Language, Classrooms and Computers*. London: Routledge, 1993.

Underwood, J. and Underwood, G. *Computers and Learning*. Basil Blackwell, Oxford 1990.

Underwood, G. 'Collaboration and problem solving: Gender differences and the quality of discussion'. In Underwood J. (ed) *Computer Based Learning*. David Fulton Publishers: London, 1994.

Children in Control

Carol Fine and Mary Lou Thornbury

> The idea that mathematics is to do with the body has inspired me to use the computer as a medium to allow children to put their bodies back into mathematics.
>
> Papert (1993)

The work discussed in this chapter is mainly concerned with control technology in the nursery and early years of schooling. We examine the use of a controllable robot with children from 4 to 8 years old and also the use with 7-year-olds of the computer and peripherals to control lights, buzzers and switches. We refer also to the implications for teaching student teachers and teachers on in-service training (INSET). We consider whether the introduction of control can promote new ways of learning and thinking (Papert, 1980) and whether children using control can show evidence of the high order thinking and learning that they have not always been able to demonstrate within the traditional curriculum (Stewart, 1985).

The children in the nursery were using a robust robot called 'Pip'[1]. Their theme for the half term was the family life of Mr and Mrs Wolf. The Pips were dressed up with masks and the first day was divided between experimenting with the robot and transforming it into a character by dressing it up. For a few days different groups explored a route to the shops or a route to pick up Baby Wolf from school.

Bruner (1996) writes that 'infants seem to need self-initiated activity'. In this case, after the initial introduction of the robot, the children took ownership of the Pip activities, using two Pips in a game of chase or a race. Cy Roden (1997), in an article on problem-solving strategies, warns that 'children can become so preoccupied with managing and manipulating the resources that practice or self-directed play takes precedence over problem-solving'. Actually for these nursery children the 'practice' of the Pip races was the purpose. They covered the full taxonomy of skills identified by Val Warren (1992), spatial awareness (including the language of position and direction), recognition of numerals (including zero), the stages of measurement, recall of the 'grammar' or sequence of commands and development of logical thinking.

Much number learning was involved and the identification of numerals. The children counted along with the robot movement to get the feel of the numbers. As they became more confident they experimented with the robot exploring the geography of the whole nursery classroom using a random input of numbers. Finally the selection was less random and the children experimented with three figure numbers, a development which supports the observations of Gay Vaughan (1997) who, in her research with rising fives, found them willing and able to use 'double digit numbers and, where appropriate, hundreds and thousands'.

Although the children in the nursery asked about the left/right arrows to make the robot turn they were never very interested in using them. Aware of the danger of pushing against an unprogrammed motor they gently lifted Pip in the way you see children lifting cats so that the robot could go in a new direction.

The learning they engaged in here can be described as play, investigative learning, self-directed discovery learning or problem solving. These strategies are sometimes given low status in our society but it is our contention that the thinking and intellectual demands required to tackle such work require high level thought.

Traditional English primary education has not been concerned with training learners in the skills of programming. Schools have enabled children to control their own actions, to draw, paint, make, move, add, subtract, write, look up information, switch on batteries or lights, but there have not been the means to allow children to control events outside themselves through using instructions. To give instructions in order to make something happen requires a goal, a plan and a language. As Seymour Papert (1980) points out, until relatively recently our culture has not given children opportunities to think about problems systematically and our culture has been 'relatively poor in models of systematic procedures'. Control with a computer became possible in the early 1980s with the work of Seymour Papert and his programming language, LOGO. Papert, in giving the world of education LOGO, gave children 'an object to think with' (ibid., p11). The floor turtle and the screen turtle serve 'no other purpose than of being good to program and good to think with'.

The controllable robots used in the nursery were Pixie and Pip. These controllable robots look like cuboids on wheels and they are given instructions through a keypad on the top of the cuboid. They are not linked to a computer but they allow the user to program one command, or a sequence of instructions. The instructions are instructions of movement (forward, backward, left, right) and measurement (the amount the user wishes the robot to move). The actions of these robots correspond to working with the programming language LOGO in the earliest stages, in direct mode.

The role of each robot in promoting the children's experimentation was unique. Unlike any other toy or construction game in the nursery the children found themselves defining a goal or purpose in order to 'play!' with the robot. Successive groups of four nursery children working in pairs identified the place they wanted the robot to reach and they commanded it to move. Also the robot was given a role, a personality and a story as they played. The children were taking control of their own learning.

Papert (1980) discusses how children can use LOGO to 'teach the computer how to think'. Tina seemed to feel the heady experience and excitement of thinking which Papert described. A cerebral-palsied 5-year-old, she spent much of her time in the nursery by the sand, feeding people who came to her 'cafe'. Her access to many of the activities in the nursery was limited by mobility. Nor was her fine manipulation sufficient for her to work the Pip robots with their soft keypad. The new Pixie robot had raised hard keys and was small enough to use on a table top. Sitting in her wheelchair at the table she could settle with her finger on a raised key, count the number of key presses and then press 'Go'. The second time she set it going it retained the first instructions and went so far across the table top that she couldn't reach it. So we showed her the 'CM' button and told her that it meant 'Cancel Memory'.

It might have been a mistake to use a complicated phrase when the tendency among adults was to speak to Tina in short words that were easier for her to articulate in response. However Tina was undaunted; whenever the Pixie had performed a set task she would cry loudly 'cancel memory,' and press the CM button. The words corresponded to the meaning and to the importance of the action, or as Bruner (1966) maintains, it is 'the use of language as an instrument of thinking that matters, its internalisation...' Tina could instruct her Pixie to go forward, flash lights, and so on. She was able to control its wandering.

One aspect of the Pip 'play' in the nursery was the relative speed with which the children learnt the grammar of a robot command: the algorithm that would ensure control. The command <**forwards**> or <**backwards**>, on the right of the key pad had to be followed by a number, the measure of that forward or backward movement, then by GO. The children learnt to press CM if they did not want a string of moves. It took a little longer for them to be confident about allowing the robot to perform two moves or even more. By the third or fourth session the children were able to program a sequence of instructions instead of separate movements. They eagerly checked the program they had planned against Pip's movements.

Wood and Attfield (1996) write 'It must be remembered that young children are learning to learn and learning about themselves as learner'. Observing pre-school children working with Pip showed us how, as Papert claims, even at this young age the children were in control of their own learning and thinking. The programmable robot enabled the children to work in a learning environment where they were setting and pursuing their own learning aims.

Our observations of children using Pip had begun in a Year 3 classroom. First impressions had been of the anthropomorphism which the children demonstrated. Children do this all the time but it emerged as a powerful emotional and intellectual link in exactly the way Papert describes; they put their bodies into the mathematics. Val Warren also refers to such anthropomorphism in the title of her article 'We called our Roamer "John".' The Year 3 children in this observation gave the robot a name, gave it a task and endowed it with skills and intention. They were able to do this because the movement of the robot corresponded to the physical movement of their own bodies; they could act out the angles by turning through them. This 'physical geometry' is located in the body and in its relationship to space. This identification was the basis of their stories.

A class of Year 3 children were asked to make maps and pathways for Pip to explore, as a part of their Geography programme. First they made a trial map. The teachers had thought that this would be a difficult stage, that the maps would be unrealistically curvy, or so detailed that the black box could not negotiate the route. However, the trial maps were simple and neat, with little elaborations such as notes, circles or flashing lights to indicate the tasks set for Pip on the way. Each map was transferred to an A1 sheet. The children worked in self-chosen groups of two or three. Pip needed a fairly wide road of 150mm and if the robot was not to be penalised it had to negotiate turns without going over the edge. At this stage the map was trialled.

The most striking aspect of this task was the imagination that went into composing a scenario for Pip's adventure . The rather dull black, rectangular shape of the robot meant that no identification was too bizarre; Pip could be any sort of adventurer. Was he / she to explore a delicately drawn jungle, beset by tigers, look for treasure on a desert island or simply move to a lovely new house in the country? One statemented boy, who preferred to work on his own, made Pip into a car; he made streets and a route to a car park and embellished the route with a pedestrian crossing and a compulsory stop. A very quiet girl, together with a rather boisterous boy, worked on a space route for Pip: the girl's brother in Year 5 was 'doing' a project on the planets so he was called in to advise on the order of the planets en route to Jupiter and to give help in drawing them.

Working with control robots as part of the Geography curriculum with this class led to stories of Pip's adventures. All the excitement of planning the route translated into stories that were longer than anything they had written before. The need to follow a route imposed a sequence on the stories, giving them a coherent beginning, middle and end. What began with the narrow brief 'give a set of commands that produce a variety of outcomes, and describe the effects of their actions' (PoS 3 Controlling and modelling KS1 National Curriculum for Information Technology) resulted in a complete study involving linguistic and artistic skills.

In recent years there has been a renewed pedagogic interest in story telling. Gudmundsdottir (1991) writes 'curriculum stories generate numerous shorter stories that teachers tell as explanations or illustrations of a larger idea. These stories are part of the narrative way of knowing that is basic to the ways in which human beings understand the world and communicate that understanding to others.' Narrative ways of knowing are not just part of the English or History curriculum; they are also the ways we first recount technological learning; we tell what we did. If we cannot tell it then it is unlikely that it has been understood[2].

The building blocks of narrative are important in the development of thinking. Carol Fox (1989) suggests that 'argument is implicit in narrative discourse itself'. These building blocks 'and then', 'but when', 'if' or 'because' are those of argument, and by sorting the events of a story, in terms of cause and effect, motivation or character, children use the tools that they will later command in the more abstract argument demanded of mature learners.

As the children's Pip stories show, technology is not outside this schema. The 'why' and 'how' of making and remaking and the debugging of a program in control or modelling, conform to algorithmic structures that are also expressed in the ordering of events in stories. Saddiq wanted Pip to cross a chasm; 'Can he jump?' he asked. When he had worked out the answer he decided that the chasm would need a bridge. Whether this was 'debugging' or 're-plotting' Saddiq was both making stories and using algorithms.

What is it about Control Technology and controlling a robot that links with the structures of story-telling? The Year 3 Pip technology task brought to the story-telling a sense of purpose with a sufficient structure and with real constraints. The sense of purpose derived from the 'where'. Where did the character in the story want to go? Linked to that was the 'why', the purpose of the journey/adventure. The characterisation of the robot was derived from these purposes. Bruner (1986) suggests that one of the ways we have of 'knowing' is the narrative that looks for connections between events. The children looked for a 'beginning' and a purpose for their story. To find a treasure they needed a map; to reach the moon they needed a rocket, to go to the mosque they began from home in their best clothes; on a trip to the park they first called on their friends to accompany them.

The significant event which provides climax, impetus and suspense will arise also and it is interesting how different children arrive at this element. Two examples from a Year 2 class illustrate this:

Alfie and Nathan had decided that their treasure would be found in a pyramid; they knew about the variety of terrain that would be encountered, providing the element of adventure. They began the rough plan for the story by drawing their first map (Figure 1). There was a little picture of Hercules digging in a garden, getting into a car, boarding a ship, landing in the jungle, negotiating the crossing of a river with canoe and dangerous crocodiles and... At this point they wanted to climb mountains but realised that their story would never end if the chain of narrative were not rounded to a conclusion. They had in fact almost forgotten where they were going but went back and 'read' the map. In their revision the jungle river emerged into a desert terrain in which stood the pyramid and one last hazard, a snake guarding the treasure. The arrival at the pyramid had required a heightened sense of danger, which the snake provided; so the story

Figure 1 This is Nathan and Alfie's first try at a map, before they realise that they cannot fit any more on

reached a climax. Their fertile imaginations had also been reined in without the teacher calling time.

At first Alfie and Nathan had mapped their story out like a cartoon; when the narrative reached the right-hand edge of the page it began again on the left. When they transferred the story to their final large paper they found that the robot could not negotiate the 'cartoon' plan. They corrected the layout by making the robot weave like a snake across the page. As Wood also points out, the ability to regulate one's activity - to recognise that the first impulse is not always correct - is an intellectual achievement.

Victoria and Sarah wanted their Robot, Kitty, to go to the moon. The two girls worked separately; they divided up the drawing of the map and the story writing but carried on their conversation to bridge the gap. The story began to go into all the detail of a plane journey, looking out of windows, having meals and sleeping. Victoria was drawing the map (Figure 2) with a rocket pad at one corner and the moon in another with a space between. She suggested that 'Kitty' needed to do more so they drew the Earth in the middle of the map for the rocket to circle, 'when she looked down she saw the green of the trees and the blue of the sea'. As

Figure 2

Wood (1988) writes, 'One of Vygotsky's theoretical arguments is that 'self-regulation' is discovered and perfected in the course of social and instructional interactions' . At this stage Sarah, the writer, stopped looking up every word in the dictionary and became lyrical. The need for a climactic event became evident; the rocket veered off course, approached the sun and was in danger of burning up. Kitty escaped however and the story ended cryptically with, 'they tried the moon'. The map-making extended the narrative so the algorithm needed further refining when the robot and map were brought together. The map had provided the structure for the programming of the robot.

There were also challenges in the programming. Both pairs of children quickly recognised that the measurement of distance was in centimetres and were also able to establish by trial and error what degree of turn they needed. The pyramid adventure relied more on right angle turns than did the moon adventure but both used the little flashing light and some musical sounds for the snake and the arrival on the moon. Finally both groups wrote out their story and performed it to the class. So the project ended with a public story-telling with a performing robot as a prop.

With the other Year 2 and 3 classes where the theme was more geographical we looked at various story books with maps in, but in particular at the beautifully illustrated *House Cat* by Helen Cooper (1993). We also played with the robot in the middle of the large circle of children who experimented with distance and angle till they were able to program Pip to turn and travel to their friends across the central space.

There were many ways in which the children we observed arrived at the correct numbers for the angles. To begin with they tried left or right 2 or 3 and the robot twitched. When they tried bigger numbers it finally made a substantial turn. One thing they learned is that the turn and the subsequent travelling are not the same. So they learned what is very difficult to learn on paper; that angle is the measure of turn and not of distance.

Despite the present description of angle in the National Curriculum (PoS 3b understand angle as a measure of turn and recognise quarter turns and half turns, *e.g. giving instructions for rotating a programmable toy*) some children emerge from their Geometry teaching with a belief that angles are 'rays'. We found that if children study the circle they learn without confusion. They can associate angles with the numbers on a clock face or with the points of a compass. Turning through 360 degrees they use the physical geometry which matches their own body movement. Papert (1993) describes children gyrating through the degrees and this is the experience of every primary teacher who has worked with robots or LOGO, '...when you let yourself go you will find that there is a richer source of mathematical knowledge in your body than in classroom textbooks'.

Our work using Pip with Year 2 and Year 3 children revealed how the children used language in identifying their goals and in planning their solutions. This led us to the activities with nursery children which we have described. We followed up by observing children using Pip in reception and year one. The observations, together with the work on computer control, led to the development of a plan for progression in control skills which is at the end of this chapter. Such a progression is important and the 1996 OFSTED report, 'Subjects and Standards' (p.11), identified as a key issue that headteachers and co-ordinators 'ensure that IT provision is coherent, complete and progressive'.

Computer control

Computer control enables the user to switch on lights, buzzers and motors so that something is clearly seen or heard to happen. There is an effect. A switch can be pressed, a magnet used, some mercury tilted, a temperature reached and this can cause the lights, buzzers or motors to work. The computer control, as with the controllable robot, provides a concrete experience and offers an instant response. If we accept 'the centrality of purposive activity in learning' (Wood, 1988) then computer control is a wonderful tool with which to engage the learner. As Loveless (1995) claims 'control technology offers the opportunity for an investigative approach to solving problems and a development in the complexity of giving and modifying instructions. It also offers the chance for children to engage in processes which they can relate to the real world'.

Unfortunately the literature documents only isolated primary classroom examples where control has been used. In the early 1980s the government-funded MEP (Micro Electronics Programme) produced some useful case studies of teachers using control as part of their classroom curriculum. Between 1985 and 1987 the ILEA (Inner London Education Authority) carried out a small control project. Debbie Johnson (1990) monitoring the ILEA project, noted the lack of expertise in design and technology, the high cost of hardware and the large commitment of teacher time which all mediated against the use of control being taken up in primary schools. However she also noted that many of the teachers taking part in the project were convinced that control was a logical and 'concrete' way forward in using programming where the problems were relevant and meaningful to the children.

During the 1990s the IT literature has been increasingly directed by the needs and the requirements of the IT National Curriculum. Chris Drage (1996) illustrating a tendency to take a rather narrow approach to control writes 'I cannot think of a more enjoyable and motivating way to introduce the basics of control technology to the whole class and tick off those aspects of the programmes of study to boot!'. Other articles indicate that control is seldom being used in primary classrooms. Recently *Microscope* (1996), in the introduction to a special edition on control, began 'Using a computer for "control" can prove a daunting task'. Another more recent article by Ron Allen begins 'Well ...What is this control and what has it got to do with technology?'.

Our experience in teacher training also indicates that there is very little LOGO or control taking place in London primary schools. Students rarely find LOGO being used in the classrooms where they practise their teaching and in only one case in 10 years has control been seen. The National Curriculum for IT in 1988 and its revision in 1995 seems to have made little difference to this student experience. The most recent OFSTED review of Information Technology (1993/4) found 'Pupils are not confident with control which is often absent after key stage 1' and the 1994/5 OFSTED report on 'Subjects and Standards' found 'more work takes place with robots and toys........ Even so, over one-third of schools do not meet the requirements of the National Curriculum'.

It is against this background that we decided to work with a Year 2 class (6 and 7-year-olds in an inner city London school) to introduce computer control. We took a buffer box and some inputs and outputs into the school as they did not have their own control equipment. The class had some experience using the controllable robot, Pip, but they had not used LOGO. There was access to two computers which were mainly used for word processing. The children were also confident in using a variety of 'story' CD-Roms, primarily to help their reading and enjoyment of a story.

The class was engaged on a topic on pirates and it was suggested that they make pirate masks which might light up, make a sound or even move. The methodology was straightforward and one afternoon during the story period the whole class was shown how the computer could be commanded to switch on a light, a buzzer and a motor. The class was shown a mask which some other children had made as an example. These children were then asked to design a pirate mask and to indicate, in their design, the bits of the mask which they wished the computer to control.

The opportunity to design, make and control was an irresistible combination and the whole class was enthused. Bruner (1966) suggests the teacher needs to provide a learning environment where the children are involved in self-directed learning, that the teacher should enlist 'curiosity, a desire for competence, aspiration to emulate a model and a deep-sensed commitment to the web of social reciprocity'. It seemed that the mask control task met Bruner's conditions. The work took place over two weeks as part of the other classroom activities. As the children completed their masks they took them to the computer to hook up to the various lights and switches which they had selected. One computer was dedicated to control over the two weeks. This meant that even when there was no-one ready to attach their mask the children could type in commands to light the lights, sound the buzzers and turn on the motors. The children took full advantage of the opportunity to practise and quickly learned to type in the commands they needed, either *switch on* followed by a number *(from 1 to 8)* or *motor A (B, C, or D) on*.. The fact that a special and accurate language was required to instruct the computer was quickly absorbed by every child in this class which was balanced in its range of ability. The children found it entertaining to plug the lights and buzzers into the variously numbered outputs, making sure that they typed the correct number in order to make them switch on or switch off. There was one child in the class who was just beginning to learn English but easily learned the syntax necessary to switch things on. Papert (1993) condemns the 'faddish idea that what children learn should be made relevant' and he cites examples of asking children to work with sums by giving them pretend shopping problems. We hoped that the children did not think they were playing 'a silly game' with this task and the seriousness with which they designed their masks suggested that they were exploiting and consolidating much of the learning which had already taken place during their pirate topic. Many of the children selected a named historical pirate to illustrate and they were all careful in the details they used such as hats, earrings and gender. Some chose the female pirates they had learned about. The context for using control is as important for learning as the context for any other learning activity (Claxton, 1984).

The designs for the masks indicated which feature/s the children wanted to control. A surprising aspect of this project was the way in which most of the children displayed a realistic grasp of their own limitations. The younger and the less able children attempted simple designs with one action. They simply aimed for two eyes to light up or one nose to light up. The more confident and more able children included more complex actions such as eyes lighting up *and* the mouth making a sound. Two children included the use of a motor to make a hat spin and a nose rotate. These children were not the most capable in the class but they were able to tackle the programming task they had set up for themselves. We were able to observe a class of children displaying a range of different developmental stages. It seemed that the children had a perceptive insight into their own abilities and they targeted their objectives or goals within their potential grasp.

We also compared this control activity to the stories children of the same age had been making with the controllable robot, Pip. It occurred to us that both activities involved the building of a simple narrative or sequence. In the case of control the children chose which lights and/or buzzers to use and in which order each event should take place. This corresponded to decisions about events in the plot of the Pip stories.

As the children completed their masks they came to the computer to connect the outputs they had planned for. They always made reference to their original design. At the computer it was clear that physically attaching lights, as well as typing in the relevant commands, was much more easily done with a friend. Vygotsky (1934 tr.1986) writes that 'thought development is determined by language, i.e. by the linguistic tools of thought and by the sociocultural experience of the child'. The interdependency of language and thought could be observed as the children sat at the computer working out how to control the switches for their masks. The discussions were almost always on task and much thinking out loud, listening and problem solving took place. There was a sense of satisfaction with a job well done as each mask 'worked'.

Two children wanted their mask to work in a more sophisticated manner. Both were girls. They wanted their eyes to light up and a motor to turn a 'bee' on the pirate's hat. This meant that we had to teach the girls how to make a small program or procedure in order to ensure that several events could take place at once. The girls accepted this extra bit of programming happily and were able to understand it and use it with their masks. The discussions supported Bruner's (1966) assertion that 'knowing is a process not a product'. At this stage one or two of the children began some 'what if' questions such as 'what would happen if a buzzer was added to the procedure?' The teacher role was important here, to elicit predictions and to scaffold experimentation. Although other children were interested in the possibility of making a program none of them asked whether it could be used with their own mask.

In identifying what the children learned during their introduction to computer control it is important to note that the performance of the task relied on the children setting their own goals or purposes. They designed their mask and decided how they would like it to work. In order to design their mask they needed some understanding of how the computer could help them control the action of their masks. The children learned that the computer can control one event, or in some cases a sequence of events. The children learned the particular syntax or language which they were required to use in order to control the computer and this did not seem to present any problems. The children programmed the switches and in some cases worked out why an event they had planned and predicted was not working.

As Papert (1980) writes 'the question to ask about the program is not whether it is right or wrong, but if it is fixable'. The use of computer control in the context that has been described was both challenging and accessible for 6 and 7-year-olds. The concrete nature of the task enabled the children to pose their own problems and solve them by learning to control the computer by using a new language. This type of activity may well be useful in laying the foundation for future and more complex programming activities.

The children (aged 6 and 7 years) were working at Key stage 1 giving 'direct signals or commands that produce a variety of outcomes, and describe the effects of their actions' (National Curriculum for IT, POS 3b) and also at level 3; 'They understand how to control equipment to achieve specific outcomes by giving a series of instructions'. It is undeniable that at Key stage 1 and possibly at Key stage 2 as well, the use of a controllable robot and the program LOGO can satisfy the N.C. IT requirements in control. However the potential for experimentation and investigation offered by the use of buffer boxes and switches cannot be replicated.

What makes computer control different from other IT applications? The user is impelled to experiment with, use and understand commands in order to achieve a goal. The sequence of cause and effect is mediated through language. We found that language and dialogue, between children and between teachers and children, mediated the learning. Vygotsky (1978) writes 'children solve practical tasks with the help of their speech, as well as their eyes and hands. This unity of perception, speech and action, which ultimately produces internalisation of the visual field, constitutes the central subject matter for any analysis of the origin of uniquely human forms of behaviour.'

Control is a creative activity requiring the technological approach of trial, error, evaluation and testing. Interestingly control activities are often identified primarily as mathematical activities, the contribution of both language and some sort of scientific methodology being ignored or sidelined.

Many of the examples in this chapter are types of activities often described as 'problem-solving'. There is a difficulty with such nomenclature. Anita Straker (1987) writes inspirationally about the use of computers for problem solving, quoting the use of the term in the Cockcroft Report and describing the negative experiences of children in this area until the advent of the computer: 'Problem solving is like cooking: it is something you learn about and become more skilled in by actually doing.'

The learning described here is not a 'problem' but here we do meet the problem that our language does not adequately describe this sort of learning. Papert calls it a 'hobby', many describe it as 'play' but because the language does not equate those words with serious engagement the activity becomes problematised even when we want the children to be exploring and discovering. Look in the thesaurus for similes for 'problem' and you find puzzle, dilemma, doubt, issue, uncertainty, enigma, question, and riddle, all negative words. Most people, and children especially, fail to remember the negative. This sort of postulation disempowers them; the teacher must have the answer. So we have a recipe for teacher control of learning. Pupil autonomy comes when they have a task, job, assignment or even a chore or duty which they can fulfil.

The word 'problem' does not have a negative value only for pupils; language enslaves the teachers too and misleads them into thinking that there must always be known and right answers in the classroom. This was drawn to our attention when a research student was working with her class of 8-year-olds on a control activity. She asked them how they were going to solve the 'problem' they had set themselves and they began to look for something wrong with their plan: 'The children did not approach problem solving with a negative attitude; they regarded the activity as a challenge, a form of critical thinking. Perhaps the problem is only a problem to the adults.' (Michelle Reilly, 1994)

At the end of this chapter are two examples of control challenges that have been given to teacher training students. They are tasks which students have tackled at their own level but which could be tackled by Year 6 children if the school had a scheme of work for measurement and control which began in the reception class. These tasks make explicit how narrative is involved in enabling learners to program. In our examples students were required to tell the story of a ghost train or of Medusa in order to help them sequence and refine their programming.

The Pip stories, the mask making and the student tasks teach investigation and problem solving through narrative. Bruner (1986) identifies two fundamental ways of knowing. The first is the 'paradigmatic' way, concerned with scientific method. The second is the 'narrative' way which looks to connections between events. Our work shows children and adults combining these ways of knowing. The narrative way gives access to the paradigmatic way of learning[3]. The Pip stories provided a framework in which the children could build programs, the masks gave the context for children to instruct the computer and the student tasks also performed this scaffolding function.

We began working with control with children in the early years in order to gain some insight into how such work might change children's thinking in the classroom context. We found that both the controllable robot and the use of computer control using switches offer children the unique possibility of controlling their own learning.

Two examples of control tasks

Ghost Train

A School of Teaching Studies is, as usual, short of funds. The Head of School feels that a 'funday' might trawl in some cash. The big crowd puller for the event is to be an up to the minute, technologically sophisticated GHOST TRAIN. A GHOST TRAIN is a traditional English fairground ride which usually consists of a train which takes passengers through a dark and frightening tunnel. At intervals during this ride frightening and sudden events happen to cause screaming, faintness and terror and great fun and entertainment is had by all!! The Head of School turns to the top design students, and asks them to come up with something different.

The task

The members of course *X* will work in design groups of two or three. By (the allotted time) each group will have designed a frightening event on one section of the ghost train. This event will be ***controlled*** by the computer. You should aim to have a working model of this event to demonstrate to the rest of the course members.

Perseus and Medusa

King Polydectes sent Perseus to cut off the head of Medusa the Gorgon. On a rocky, dreary, barren Island lived three fierce sisters who were called the Gorgons. These fierce women had: faces like women, eagles' wings with glittering golden feathers, scales of brass and iron, claws of brass and great fierce looking tusks which must have combined strangely with their human faces. Worst of all, instead of hair, their heads were covered with venomous snakes that were always twisting themselves about and putting out their forked tongues ready to bite anything which came within reach. Medusa was the most frightful one of all because her face was so terrible that anyone who looked at it was immediately changed into stone.

Perseus was helped by several of the gods. Minerva (Athene) gave Perseus her own bright shield. She said that when he was ready to strike off the Medusa's head he must look at that terrible face only in the shield. Mercury lent Perseus his own crooked sword which was the only sword which would cut through the Gorgon's scales. The nymphs of the Garden of Hesperides brought Perseus a pair of winged sandals, Pluto's helmet which made the wearer invisible, and a magic pouch in which he could safely carry the head of Medusa.

The task

To make a working model which illustrates an aspect of this story. The model must include at least three events which are controlled by the computer. Some ideas you may take into account are:

- Perseus needed to get near enough to the Gorgons to cut off the head of Medusa. How could the Gorgons protect themselves?
- He had to do this at night whilst the Gorgons slept. How could the Gorgons prevent this?
- How could the other Gorgons prevent intruders escaping once they had entered their territory?

Notes

[1] This paragraph refers to work with Pips and Pixies. A Pip is a programmable robot, a rectangular black box with keys in a pressure pad set into the top for control and powered by a rechargeable battery, produced by Swallow Systems of High Wycombe, Buckinghamshire. A Pixie from the same firm is a smaller tabletop robot with raised buttons for control. A robot produced by Valiant Products is called the Roamer; it is a different shape but operates by the same Logo-type instructions.

[2] The application of IT to the English curriculum is not simply confined to Strand 1 in the IT National Curriculum, Communicating and Handling Information. Strand 2, which includes the use of control, has many implications for the use of IT in English (or the use of English in IT).

[3] There is debate which distinguishes between fictional narrative and other narrative, e.g. historical, a discussion which is outside the scope of this article. Programming itself could be seen as another narrative form. An author who takes this discussion further is Marie-Laure Ryan in her book *Possible Worlds, Artificial Intelligence and Narrative Theory,* (1991) Indiana University Press.

References

Allen, Ron. 'Computer Control - a cross-curricular approach to teaching'. *Microscope* **49** (1997).

Bruner, J. *Toward a Theory of Instruction.* Harvard University Press: Cambridge, Mass, 1966.

Bruner, J. *Actual Minds, Possible Worlds.* Harvard University Press: Cambridge, Mass, 1986.

Bruner, J. 'What we have learned about early learning'. *The European Early Childhood Education Research Journal* **4** (1) (1996).

Claxton, G. *Live and Learn: an Introduction to the psychology of growth and change in everyday life.* Harper and Row Publishers, 1984.

Cooper, Helen. *The House Cat.* Scholastic Publications, 1993.

Drage, Chris. *Educational Computing and Technology.* (June/July 1996).

Fox, Carol. 'Divine dialogues: the role of argument in the narrative discourse of a five-year-old storyteller.' In Andrews, R. (ed) *Narrative and Argument* . Open University Press, 1989.

Gudmundsdottir, Sigrun. Story-maker, story-teller: narrative structures in curriculum. *Journal of Curriculum Studies* **23** (3) (1991).

Johnson, Debbie. *Control Technology in the Primary School.* ILECAS, 1990.

Loveless, Avril *The Role of I.T.: Practical issues for the primary teacher* Cassell, 1995.

MEP. *Control and Technology in a Primary School,* 1985.

OFSTED *Subjects and Standards, 1994/5 Report,* HMSO: London, 1996.

OFSTED *Information Technology, A Review of Inspection Findings.* HMSO: London, 1993-4.

Papert, S. *Mindstorms Children, Computers and Powerful Ideas.* The Harvester Press: Brighton, 1980.

Papert, S. *The Children's Machine.* Basic Books: New York, 1993.

Reilly, Michelle. Honours research, Teaching Studies, University of North London, 1994.

Roden, Cy. 'Young children's problem-solving in design and technology: towards a taxonomy of strategies,' *Journal of Design and Technology Education* **2** (1) (1997).

Stewart, *J. CAL* 1985

Straker, Anita. 'Thinking things out: solving problems with a micro.' In Fisher, R. (ed) *Problem-solving in Primary Schools*. Basil Blackwell, 1987.

Urwin, Bill. Switch-on (an introduction to Control). *Microscope Software Special.* published by Newman College with MAPE (the Association for Micros in Primary Education), 1996.

Vaughan, Gay. ' Number Education for Very Young Children: Can IT change the nature of Early Years mathematics education?' In Somekh, B. & Davis, N. (eds) *Using Information Technology Effectively in Teaching and Learning* . Routledge: London, 1997.

Vygotsky, L.S. *Mind in Society* . Harvard University Press, 1978.

Vygotsky, L.S. *Thought and Language,* Ed and tr. Alex Kozulin. MIT Press: Cambridge, MA, 1986.

Warren, Val. We called our Roamer 'John' in *Strategies* **2** (3) (1992).

Wood, D. *How Children Think and Learn.* Blackwell: Oxford, 1988.

Wood, E. & Attfield, J. *Play, Learning and the Early Childhood Curriculum.* Paul Chapman Publishing, 1996.

Appendix: Progression in controlling and monitoring

This list is meant to complement and reinforce the work taking place throughout the primary curriculum. It represents a minimum IT competence in order to guide and support the class teacher.

Year	Skills and concepts	Useful Software Application/s
Nursery	Moving in space, forward and backward directions	games in the playground and the gym
	Introducing direction arrows	
Reception	Giving verbal commands to each other to move in different directions, forward, backward, left and right. Draw on the child's well established knowledge of 'body geometry' as a starting point.	include turning in hall games
Year 1 **and** **Year 2**	Giving forward commands to a controllable robot Giving backward commands to a controllable robot Experimenting with big numbers	PIP PIXIE ROAMER
	Giving directions for turning a controllable robot. Introducing Roamer mats and PIP mats for tasks to predict steps and distance	hall or gym work
	An understanding that a **command** tells the turtle **how to move** (fd bk lt rt). To tell it how far to go **the command must be followed by a number**	
Year 3	Use a *short* series of instructions to direct a controllable robot (e.g. around an object, to post a letter or through a very simple maze). Rehearse such work with robot playing tasks in the gym.	hall or gym work ROAMER PIP PIXIE
	Introduce LOGO in direct mode (one command at a time) Introduce abbreviations fd, bk, lt, rt CS (clear screen), CT (centre turtle)	LOGO
	Encourage children to write down their instructions so that they can program the robot or screen turtle to carry out a *short* sequence of instructions in one go. Encourage the testing of these instructions.	

Year	Activities	Software
Year 3 cont'd.	Introduce Computer Control by giving experience in controlling outputs as part of a curriculum or topic based task. Discuss examples of the use of Computer Control outside the classroom	SMARTMOVE COCO LEGO DACTA
Year 4	Use sentence and syntax cards supplied with Roamer or make similar ones for other robots	
	Program a controllable robot with a more complex sequence of instructions which will be stored in the memory. Encourage children to write down their instructions. Introduce the use of repeats.	PIP PIXIE ROAMER
	Continue to use LOGO in direct mode. Encourage using sequences of instructions to direct the screen turtle, perhaps to draw simple shapes. Children should write down their instructions and test them. Introduce the use of repeats. Introduce lift and drop commands	LOGO
	Extend the use of Computer Control for controlling more than one output. Introduction of simple procedures and the use of repeats for flashing lights etc. Continue to use Control in the context of the classroom curriculum or topic Reinforce earlier discussions on Computer Control outside the classroom, in the home in commerce and in industry	SMARTMOVE COCO LEGO DACTA
Year 5	Introducing the idea that you can teach LOGO a new word by writing a procedure (e.g. teaching LOGO to SQUARE) Introducing the process of debugging LOGO procedures, possibly by playing turtle.	LOGO
	Using Computer Control introduce the use of inputs and the discussion of "what if?" questions Continue to use control in the context of the classroom curriculum or topic. Discuss the place of trial and error in design. Link these discussions and classroom control activities to the approaches of science and engineering to the problem solving process.	SMARTMOVE COCO LEGO DACTA

Year 6	Introducing variables in LOGO so that children can draw shapes and objects of different sizes	LOGO
	Continue to "play turtle" in the gym or hall to rehearse these skills	
	Introducing the idea of nesting procedures in LOGO (e.g. a procedure for rocket may be top body stick)	
	Consolidation of the use of Computer Control using procedures. The use of subprocedures to encourage more elegant problem solving and programming.	SMARTMOVE COCO LEGO DACTA
	The design of problems or tasks should enable the children to tackle them with increasing independence. Children should be working in groups and be encouraged to analyse the group approach to problem solving.	
	Discussions should continue to compare the control tasks in the classroom to similar situations outside the classroom.	

New Ways of Telling: Multimedia Authoring in the Classroom

Bill O'Neill

Introduction

As the digital revolution transforms the educational landscape, insightful judgements are called for about educational values and the best uses of available technologies to achieve goals which derive from or rest upon those values. This digital world will increasingly require flexible and versatile thinkers capable of interacting in complex situations. The argument in this chapter is that computer - based multimedia provides educationalists with a powerful set of tools; the struggle is to learn how to make creative use of these sophisticated tools.

A new information age

This is not the first information age! Even if we do not learn from history there is some relief in knowing that others have similarly struggled to exploit available technologies to promote educational goals.

Each and every one of us relies heavily on the inventions and tools of previous generations. Ironically, however, there is frequently a reluctance to embrace the newest developments in technology; this is understandable and, perhaps as it ought to be. Our everyday, ' taken for granted technology' has stood the test of time; each and every tool was designed, in its day, to extend some aspect of human potential. Through time, with use and debate about that use, the tools have evolved. In such a manner everyday technology bears the influence of the multitude of users who have shaped and refined it. We are witnessing another iteration in this process; as users we are all party to the shaping of computer-based technologies which are sure to become the 'taken for granted' technologies of tomorrow.

Positioned both to reflect the changing values and aspirations of the culture and to exploit developments in technology, educationalists typically find themselves at the fulcrum of this powerful dynamic; a dynamic which causes a re-evaluation of the nature of education itself. This shift is especially apparent during epochs of dramatic developments in the technologies of information. It suffices to mention briefly four major epochs in information technology: speech and oral traditions, the inventions of literacy (and numeracy), the printing press, and the computer.

1) From the dawn of civilisation, story has been the carrier of information. Story tellers played a central role in early societies; in Gaelic Ireland, for example, the educated person was a poet and story teller.

2) The invention of literacy brought about a revolutionary way of thinking; the new technology allowing knowledge to be coded, stored and reflected upon. Education became defined in terms of access to this technology (physical availability and 'decoding' skills). Based on this new information technology 'data banks' were created; one of which, the magnificent library of Alexandria, in the first century BC, which is said to have housed close to three quarters of a million hand crafted texts, became the centre of a thriving information culture.

3) In the middle ages the new technology of the printing press revolutionised communication. A new information age had begun. This new mechanised technology of communication allowed wider access to the technology of literacy. Educationalists, for their part, responded by developing teaching methods aimed at furthering the goal of universal literacy.

4) Computer based communication is heralding a new information age; an age in which text, graphics, sound, video etc. is increasingly stored in computers in digital format. This digital age is again revolutionising communication and bringing about a significant shift in emphasis from the written word to world wide multimedia communications.

Engaging with the new

We can confidently assert that change is inevitable. Each successive epoch, in providing increasingly powerful technology, has served both to expand the knowledge base and to allow greater access to it. In previous epochs it is likely that reactions to the developing technologies were not dissimilar to those of today; some resisted, some converted, others debated. The historic challenge to educators is to engage in the debate in order to find ways to turn the developing technologies in the direction of and to the advantage of education.

In this newest information age some, pejoratively labelled 'schoolers' by Papert (1993), resisting change, harking back to a supposedly golden age, have used the opportunity to call for a return to traditional values. Unfortunately for them, history teaches us that there is no going back; the 'basics' which sustained our yesterdays will not nurture our tomorrows! Sensing the inevitability of change there have always been some who seek their solutions in the technology itself; in our own cycle of these events, as O'Shea & Self (1983) put it, 'Microcomputer developers are luring educational technologists back to an emphasis on devices rather than concepts'. The logical conclusion from this perspective is that the computer will eventually replace the teacher. Research, however, supports a very different conclusion, one which places the teacher at the centre of the education process. For example, in a recent review of technological innovation in education the centrality of the teacher was affirmed:

> Regardless of the focus or scope of the innovation, it appears that one type of result consistently occurs; a result that acknowledges the teacher as the key figure in the eventual success or lack of success of any computer-in-education initiative.
>
> Collis B. (1996)

Bronowski (1973) pointed out that the technological revolutions throughout the ages were all social developments in the use of cultural tools; technology cannot replace people, rather it is a system of tools devised by people to extend their power. As new tools extend choices users are caused to exercise judgement in their use; as the power of technology increases more sophisticated theoretical frameworks are required to inform these judgements.

Bednar et al. (1991) in an aptly named work, *The Society of Text: Hypertext, Hypermedia and the Social Construction of Information*, has argued that social constructivism (Vygotsky, 1962) provides a paradigm for this new digital age of world wide communication and information. They argue that learners must be provided with the available tools placed in rich problem solving situations reflective of real world contexts. Computer based digital technologies have the characteristics to fulfil this role. Bednar et al. outline some of the implications of this constructivist approach: 'In this view, learning is a constructive process in which the learner is building an internal representation of knowledge, a personal interpretation of experience .... Learning is an active process in which meaning is developed on the basis of experience.'

Social constructivism

At its centre Vygotsky's theory is a radical challenge to the individualistic learning theories of Piaget and others which have proven so influential in education. Vygotsky argues that learning is a social construction of knowledge with language at its core; that mediation, especially by a teacher, is central; that learning has an historical dimension, involving the use of cultural tools.

This is no grand theory which offers easy solutions; learning is recognised as problematic, a complex social process which places the teacher in a central mediating role. Vygotsky (1962) developed the fundamental concept of 'consciousness, or deliberate mastery' to explain the relationship between school instruction and the mental development of the child:

> The psychological pre-requirements for instruction in different school subjects are to a large extent the same; instruction in a given subject influences the development of the higher functions far beyond the confines of that particular subject. The main psychic functions involved in studying various subjects are interdependent - their common basis are consciousness and deliberate mastery, the principal contribution of the school years.

Thus it is argued that consciousness and deliberate mastery are the fundamental basics of school learning; this awareness of one's own thinking, or intellectual self control is developed in a highly social process grounded in a particular historical and social setting and makes use of the cultural tools of society, or as Bruner, in the introduction to Vygotsky's (1962) *Language and Thought*, explains 'Man, if you will, is shaped by the tools and instruments that he comes to use and neither the hand nor the mind alone can amount to much.'

From a social constructivist perspective the emerging multimedia technology offers teachers a powerful pedagogical tool-kit.

Children as multimedia authors

In exploring the use of this technology teachers are not caused to reject present practices, rather to subject them to critical review and to build on best practice. For example, it has been increasingly appreciated that, in order to develop a deep insight into literacy, children need to become part of the whole communication process. They need to become authors. To this end the word processor has proven an excellent printing press enabling children to 'publish' their own books and thus become members of the community of literacy. In a like manner, in order to develop a deep insight into computer-based digital literacy teachers and children need to become active, creative users of the technology; multimedia authorship provides new ways to make meaning (Druin and Solomon, 1996; Marcus, 1993; Monteith, 1993) and serves as an introduction to, or an apprenticeship with, the community of digital literacy.

Collis (1996) has recently argued that the major lesson learned from experience of more than 15 years of computers in education is to 'begin with teachers' own classroom problems and concerns; do not begin with the technology or its promise'.

Two consistent results have emerged from more than a decade of classroom based research (McMahon and O'Neill, 1993). In the first place teachers need to find ways to place the technology in the hands of children; and secondly teachers must develop techniques to engage and challenge children as they use these tools. In this chapter, in taking up this challenge, an approach to multimedia authoring is outlined. It will be argued that, through multimedia authoring, an environment can be created to enhance good classroom practice, incorporating four key elements in a more or less overlapping cycle of:

1) ownership
2) language
3) representation
4) reflection

Ownership

Children come to school with a vast experience of the world. Their elementary thinking is grounded in actions or as Walkerdine (1988) would have it, in the social practice of their everyday living. The strange practices of schooling (Donaldson, 1978) make little sense to many children. Even after years of schooling many children feel no ownership of school knowledge:

> Why are students not mastering what they ought to be learning? It is my belief that, until recently, those of us involved in education have not appreciated the strength of the initial conception, stereotypes, and "scripts" that students bring to their school learning nor the difficulty of refashioning or eradicating them.
>
> (Gardner, 1991)

In attempting to make sense of schooling, children 'read' events in terms of their everyday experiences; for many, schooling has no immediate significance. In attempting to create a stimulating and challenging environment teachers need to create links from everyday practices to those of schooling.

When multimedia tools are placed into the hands of children in an apprentice-like setting, they are given access to their own rich tapestry of known experiences in an assessable manner. A digital camera, for example, immediately allows the children to photograph each other; once the images are in the computer many possibilities present themselves. The children know that they own the images and have some say about the next step. Typically the discussion surrounding the editing of such images is very animated; children often have strong views and they need avenues of expression. Subsequently these images can be incorporated into a document and printed. In addition the images have become a resource to be used in a multifarious manner.

The ability to add one's own voice to the image on the screen (click on the photograph and hear the voice) adds an important new dimension. Other images or sounds, either produced directly on the screen, or taken from a source such as the Internet, can readily be incorporated. With this, the beginning of their own multimedia autobiography, they are introduced to the modern world of multimedia communication.

Through using hypertext links diverse aspects of students' work can be linked to each other, for example, a hypertext link between a picture and a story might acknowledge authorship; or a link to a child's seat in a computer map of the classroom might bring up a personal history; or even, a link to a map of the USA signifying a distant relative who emigrated to America: thus creating a sense of coherence . Thus hypertext links can create 'chains of signification' (Walkerdine, 1988) drawing relationships from the known to the increasingly distant.

Language

For Vygotsky, higher order thinking originates as actual relations between individuals in dialogue; through language our social worlds are established and sustained. Language, however, does not carry meaning in abstraction, rather language serves to signal the use made of words in specific contexts. The same words can carry different signification from one context to another and from one person to another, for example, from the context of the home to the school and between children.

The challenge for the teacher is to create a classroom context in which, through dialogue, a genuinely collaborative meaning making process can develop (Wells, 1986).

The multimedia authoring workshop can be designed to facilitate this process through a focus on conversation and reflection on language. Groups of children can become the experts in particular 'trades' such as scanning into the computer, using a digital camera, sound technician, etc. Just as these children operate as an apprentice to the teacher, the other children in their turn operate as their apprentices. Thus the process can become a dynamic social engagement. In such a conducive environment, in our experience, children are more inclined to work collaboratively; discussing and planning their projects, talking as they engage in their projects and critically evaluating the ongoing work.

What is intended here is an environment conducive to talk and to reflection on language. Donaldson (1978) argued that 'what is going to be required in our educational system is that he (the learner) should learn to turn language and thought in upon themselves. He must be able to direct his own thought process in a thoughtful manner. He must become able not just to talk but to choose what he will say, not just to interpret but to weigh possible interpretations. His conceptual system must expand in the direction of increasing ability to represent itself'.

Representation

In a recent critique on the implications of the digital age for education, Hilary McLellan (1996) argues that 'the notion of playing with representations is an important idea for educators to consider'. She develops the argument of Donald Norman (1993) who explains that representations are cognitive tools. McLellan argues that representations are important because they allow us to work with events and things absent in space and time, or for that matter, events and things that never existed, imaginary objects and concepts. Referring to the work of Gene Lave, she reminds us that 'stories are a form of representation that help people keep track of their discoveries, providing a meaningful structure for remembering what has been learned'. Out of the seamless flow of life, events are selected and sequenced, given a beginning, a middle, and an end. These emerging stories, in their telling and retelling, become our memories. Children bring such memories to school. In bridging the gap from home to school ways must be found to tap into these stories.

Writing tends to be the preferred medium of school, but not necessarily of all learners. Perhaps the most immediate advantage of multimedia authoring is that children's preferred medium can dominate. With multimedia children are able to draw their stories or projects, or to tell through photography or music. When children are given the opportunity they readily and effectively engage in the selection of appropriate modes of representation – like writing, drawing and animation – to communicate aspects of their experience.

In such an environment, text, graphics, and sound become resources readily incorporated into and expanding the possibilities of children's stories and projects. Once they have written or digitised the text, sound and graphics into the computer they begin to experience the power of hypermedia. When children are operating with digitised graphics and digitised sound they demonstrate the same willingness to cut, paste, and manipulate as they do with words on a word processor. The digitised entities can now become objects to think with. Children can begin with a drawing which once digitised might develop in a whole variety of directions. For example, one child, a reluctant learner that I worked with a few years ago, expressed a clear interest and knowledge of nature in his drawings. He was shown how to scan in his drawing of a tree and how his digital tree could be made bigger or smaller, and he quickly appreciated the possibility of animation. He eventually produced an animation which started from a seed and slowly developed into a beautiful tree, only to be felled by a woodcutter (Figure 1). Later he added a brief but apt text.

Figure 1

Learners can be helped to appreciate that each mode of representation makes its own unique contribution to the meaning making process. The case for photography, for example, is powerfully developed in the classic treatise of Berger and Mohr (1982). It is equally clear that they have much to say about the topic in hand:

> And in life, meaning is not instantaneous. Meaning is discovered in what connects, and cannot exist without development. Without a story, without an unfolding, there is no meaning. Facts, information, do not in themselves constitute meaning. Facts can be fed into a computer and become factors in a calculation. No meaning, however comes out of computers, for when we give meaning to an event, that meaning is a response, not only to the known, but also to the unknown: meaning and mystery are inseparable and neither can exist without the passing of time. Certainty may be instantaneous: doubt requires duration; meaning is born of the two. An instant photograph can only acquire meaning insofar as the viewer can read into it a duration extending beyond itself. When we find a photograph meaningful, we are lending it a past and a future.

Such a complex meaning making process engages the learner in the uses of what Howard Gardner (1993) has called 'multiple intelligences'. Gardner, who strongly criticised the notion of intelligence which can be measured by a single quotient (IQ), has proposed a model of multiple intelligences; he argues that education must support the whole range of intelligences. His goal is the development of genuine understanding which is characterised by an ability to take information and skills that have been learned in school and apply them flexibly and appropriately in a new and somewhat different setting. He goes on to argue that:

> An important symptom of an emerging understanding is the capacity to represent a problem in a number of different ways and to approach its solution from varied vantage points; a single rigid representation is unlikely to suffice.

Multimedia provides a powerful set of tools that supports multiple intelligences. As has been argued, learners can be engaged in the learning process through their preferred mode and as they begin to develop representations of their world, these representations can be talked about and eventually expressed in another medium. Through this transference from one mode of representation to another learners can begin to develop an awareness of how language and communication works.

In addition, within the hypermedia environment all aspects of the media can be linked, sequenced and re-sequenced in a complex variety of ways in order both to communicate more effectively and to view the subject from different perspectives; this virtual chain of links serves as a representation of the thinking process itself.

Reflection

Wertsch reminds us that:

> A crucial aspect of Vygotsky's understanding of human consciousness is that humans are viewed as constantly constructing their environment and their representations of this environment by engaging in various forms of activity. The process of reflection is just as much concerned with the organism's transformation of reality and representation of reality as with the reception of information.
>
> Wertsch, J. (1985)

Pedagogy involves such a process of reflection. Teaching is a deliberate and conscious activity which seeks to help the learner to direct his/her own thinking in a conscious and deliberate manner and to develop an increasing ability to represent his/her own thinking.

It is the teacher who, mindful of the three key elements of ownership, language and representation, must involve the learner in this reflection process. Donaldson (following Vygotsky) argues that this process does not happen spontaneously, that it requires the marshalling of all the resources of the culture directed to that end. Computer based multimedia is the newest and most powerful resource available to the educator. Multimedia authoring environments are rich in opportunities for teachers to engage learners in the meaning making process and to cause a reflection on how the process works; a reflection on language itself can be fostered. For example, using the simple technique of branching stories or branching accounts, the teacher can create a host of challenging contexts:

1) multiple representations

Modes of representations are different ways of telling and encourage complex rich understanding to develop. In addition a mode of representation can provide a useful framework for a project. In one investigation of a local village, for example, it was decided that a map of the area was the most appropriate focal point. The children photographed and researched prominent aspects of their village and created hyperlinks to the map. Later in the investigation, when a question relating to the history of the village was raised, it was decided to research what the village was like a 100 years ago. Using the map as the common framework, two parallel investigations could take place; one, a local study of the village as experienced by the children; the other the historical research.

2) transference from one mode to another

Decisions about the appropriate mode of representation and about the effect of shifting from one mode to the other can engage children in the discussion about communication. Children can be caused to reflect on how words and images communicate or not! Children can experiment for example, on how the combining of different sorts of music with a picture can cause a change of mood, or on the effect of text and graphics working together to communicate.

3) alternative perspectives

Locked in our own way of thinking, it is a continual struggle to appreciate alternative perspectives. Northern Ireland provides a poignant example of the need to engage in that struggle for understanding the other point of view.

In one example, a few years ago, a group of student teachers used multimedia authoring to investigate aspects of the troubles in Northern Ireland. They attempted to develop a multimedia presentation of the historical origins of the conflict. Realising that different perspectives were at the core of the problem, they faced their readers with an initial choice, either come to the problem from the eyes of a prospective planter or alternatively from the eyes of an indigenous peasant. The reader was caught up in the story from one perspective such that when both sides meet for the first time the reader came to the meeting carrying the history of one side. Representing the nature of the problem in this manner pointed to the need for understanding and compromise.

4) alternative theories can be developed; what if....?

Linear accounts assume one coherent story line; multimedia stories can allow alternative accounts to be explored simultaneously. In a multimedia story a 'what if' link can be added to allow the reader to take an alternative path. In one example, four children were co-authoring a story; at a chosen point in the story, the teacher asked each of them to suggest their own ending to the story. When all four possible endings were entered into the computer four hyperlinks offered the reader the choice to explore one of the four alternative paths. As the children began to realise that they had developed an adventure story the group began to look at the developing story from the perspective of the reader. The children found themselves engaged in a discussion of the consequences of a reader taking any one of the now growing branches; the children were analysing if, and to what extent, their story communicated.

Conclusion

In this chapter I have argued that computer-based multimedia offer teachers a pedagogical tool-kit allowing them to provide children with powerful authoring tools. Children, and learners of all ages, have developed sophisticated understandings of the world. However, although all have stories to tell, it is often difficult for them to express their ideas directly in writing. Multimedia authoring offers new choices; stories can be told through drawing or photography, or music; the learners' preferred medium can take central stage; and because our accounts are stored digitally in the computer they can be revisited, for example, to be re-told in different media. Through this transference between different media, learners can be caused to reflect on the meaning-making process itself, thus contributing to the development of 'consciousness and deliberate mastery'.

The extent, however, to which this powerful new resource will contribute to children's learning largely depends on the quality of teachers' professional judgements. Olson (1992) argues that 'the tension between the old and the new is the engine which drives critical reflection'. He points out that change involves values as well as technical issues, that values inherent in the old and the new practices are at issue. It is through dialogue that the values that underlie practice can be uncovered.

When, in the context of such critical debate, the power of the computer is put into the hands of children, the ensuing interaction between teacher, pupil and technology can greatly enhance learning.

References

Barrett, E. ' Introduction : Thought and Language in a Virtual Environment.' In Barrett, E. (ed) *The Society of Text*, MIT Press, 1989.

Berger, J. and Mohr, J. *Another way of telling*, Granta Books: London, 1982.

Bronowski, J. *The Ascent of Man*. British Broadcasting Corporation, 1973.

Bruner, J. 'Introduction in Vygotsky', (1962) *Thought and Language*, M.I.T. Press, 1961.

Collis, B. *The Internet as an Educational Innovation: Lessons from Experience with Computer Implication*. Educational Technology Nov./Dec. 1996.

Donaldson, M. *Children's Minds*. Fontana: London, 1978.

Druin, A. and Solomon, C. *Designing multimedia environments for children*. J. Wilie & Sons London, 1996.

Gardner, H. *The unschooled mind*. Fontana Press: London, 1991.

Marcus, S. 'Multimedia, Hypermedia and the teaching of English.' In Monteith, M. (ed) *Computers and Language*, (1993) 21-43. Intellect Books: Oxford, UK.

McLellan, H. *'Being Digital': Implications for Education*. Educational Technology Nov./Dec. 1996.

McMahon, H., and O'Neill W. 'Adventure Stories in HyperCard.' In Monteith, M. (ed) *Computers and Language*, (1993) 125-140. Intellect Books: Oxford, UK.

Monteith, M and Monteith, R. 'Using Hypercard in writing narratives.' In Monteith, M. (ed). *Computers and Language*, (1993) 141-156. Intellect Books: Oxford, UK.

O'Shea, T. & Self, J. *Learning and Teaching with Computers*. The Harvester Press: London, 1983.

Olson, J. *Understanding teaching*. Open University Press, 1992.

Papert, S. *Mindstorms*. The Harvester Press: London, 1980.

Vygotsky, *Thought and Language*. M.I.T. Press, 1962.

Walkerdine, V. *The mastery of reason*. Routledge: London, 1988.

Wells, G. *The Meaning Makers*. Hodder and Stoughton: London, 1986.

Wertsch, J. *Vygotsky and the social formation of mind*. Harvard University Press: Boston, MA, 1985.

Educational Multimedia: Where's the Interaction?

Noel Williams

Introduction

New technologies appear from time to time, offering solutions to all educational problems. Teachers and lecturers are, however, used to such technologies creating as many problems as they solve, and sceptical of the inevitable benefits the latest innovation is supposed to bring.

In the 1990s multimedia has been one of these 'latest technologies'. Although some teachers, remembering past promises, are sceptical, multimedia technologies may just prove to have significant and permanent educational value. Recently discussion of multimedia has been edged out by the Internet and other computer mediated communication, as well as, increasingly, the technologies of virtual reality. For my purposes, I am happy to note both the hardware and the software of these technologies are converging, such that there is little point in trying to discriminate between them. Everything on the 1997 computer can be called 'multimedia', given a particular slant, and I do not propose to discuss what is meant by multimedia merely assert that we all know a range of technologies which more or less fit the multimedia tag.

So side-stepping this familiar difficulty of how to define multimedia, we can probably agree that a significant part of that definition would be 'interactivity'. It is certainly a significant part of the hype that accompanies each new multimedia product. Educational systems are touted as flexible, interactive and user-led. In this paper I want to explore the nature of multimedia interaction. I will try to identify what educational benefit, and problems, such interactions bring with them. And I want to assert that current systems still do not fulfil the promise of educational technology, or offer the learner real interaction of educational, rather than merely instructional, benefit.

As interaction implies choice, the educational benefits of any multimedia system will depend on the choices offered, and the consequent openness or limitation of the system. So what choice do learners have?

What are the benefits of multimedia?

If there are no educational benefits to multimedia interactivity (or worse, if it creates some form of educational problem) then, like so many educational technologies before them, multimedia technologies may as well be confined to the expensive broom cupboard next to the library, for occasional use. So why would interactive computer software be of educational benefit?

Reynolds and Iwinski list the following benefits of multimedia training systems (Reynolds and Iwinski, 1996). They can:

- directly support on-the-job performance
- offer easy access to a variety of media and other resources
- offer access to information as wanted by the learner
- offer individual, self-paced instruction for individual students
- enable instructors to concentrate on instructor-related tasks, allowing them to attend specifically to students who may be having problems
- provide excellent means for simulation situations that need individualised, yet co-ordinated task performance
- use the same hardware as at home, keeping costs down.

These are benefits of different orders to different people. Those for the learner in this list all imply interactive choice:

- on the job use: a user can choose when to use it for particular purposes
- access to a variety of media and resources, as appropriate to the users' needs
- access to information as wanted, so the user selects what information she wants, as and when required
- self-paced instruction: the user chooses the rate and level of use.

These are all forms of interaction, in that they imply the user has control over a set of possible choices. However, each entails a further problem, and together they do not amount to a rich and varied educational experience, merely an increase in opportunity (and also responsibility) for the learner to access information.

The associated problems are:

1. On the job use is not truly determined by the user, but by the task. Multimedia may be more appropriate for on-the-job use (e.g. providing simulation of more immediate relevance and verisimilitude), but it does not enable the user to vary or control the task in the work environment in any new way. Consequently, the choice of access to relevant training or information is not under direct control of

the user, but is selected by the system when the task requires expertise the learner lacks.

2. Access to varied media and resources are still determined by the designer (author), not the learner. The resources available to you are those on offer, not those you might ideally require. So, whilst multimedia systems may offer more, and richer, resources, increasing the chance that there is actually a resource you want which is relevant to your needs as a learner, this is little more than any other large educational resource might offer. The onus is on the learner to make choices which are best amongst materials which have been assembled without the learner's input. The learner may not find relevant resource, because it is not there. The learner may also have a more complex task and increased cognitive load in trying to determine which choices amongst the resources are the best ones.

3. Access to information when wanted similarly requires good analysis on the part of the learners about their own needs. The bottom-line problem is represented by the learner who does not know what they need to know, or carries out flawed analysis of their own needs, and therefore accesses poor or inappropriate information. By placing the responsibility for learning on the learners' shoulders, the learning task may become harder, not easier.

So the limitations of this view of multimedia education are the obverse of its attractions: by offering more to the learner, you require more of the learner. But more resource is not necessarily better learning, because the learner still needs to be able to know what are the best choices. Furthermore a big resource is not necessarily a better resource, especially if its size is a function of 'additional irrelevance' (from the learner's viewpoint) rather than richness of relevance.

However, there is a more fundamental problem with this view of multimedia learning. Its view of interaction is limited. Interactive learning is about learners' choosing how and when to receive information which 'authorities' have assembled for them. It is entirely passive reception. Little real interaction with the materials is possible. Inter-action suggests the possibility of *action between* the learner and the information, not merely choice of how to receive it.

Why interactive learning?

So, for some systems, interaction is merely a marketing feature, an aspect of educational 'chrome', laid on top of systems which are merely information providers. Such computer-based multimedia systems cannot be rich and valuable to the learner as classroom and workroom interaction, because the types of interaction they offer are limited.

Education is predicated on interaction: either the learner interacts with a source of learning (a tutor, a book) or with reality, acquiring information and experience through that interaction. However, as we know, there are different forms of interaction with a source of learning: questioning, receptive, structured, browsing, hierarchical, serendipitous, effective, efficient, problematic, superficial, puzzling—the list is long. To say that there are different forms of interaction in learning is not to say much more than there are different approaches to education. Computer interactions offer only a subset of possible educational interactions.

Clearly, we can characterise some forms of educational experience as equivalent to some forms of interaction with computers (not merely with multimedia) and judge their value in parallel terms. Here are some examples:

Educational interaction	Computer interaction
reading a book	reading a screen
getting a different book	opening a file
turning a page	clicking 'Next Page' on screen
listening to teacher	listening to an audio sample
group problem solving	group problem solving through a computer network OR holding a dialogue with an intelligent computer
asking teacher for an answer	clicking a pop-up window

Sometimes these analogies are close, sometimes they merely seem close because, already, we have become so familiar with certain metaphors of computer use that they seem natural and intuitive. However, we might ask how natural it truly is to move a plastic box in order to move an image of an arrow onto an image of a button labelled 'Next Page', then press the physical button on the box to see the image of the button momentarily depress in order to select the next screen in a pre-programmed sequence of screens. Why does the physical button need a screen button? Why does the screen button need a label? Why does the label need a button?

Turning a page in a book (which is, again, something we lave learned to do) is a physical action without symbolic content (though it can be given such content), and has no need of metaphor, image or label to operate. Surely a more natural interface for 'Next Page' on computer would be to turn the mouse over, or to touch the bottom right corner of the screen (for Western European readers), or to physically move a piece of paper. Instead, we have to combine the iconography of page turning, with operating a video or radio, with using a pointer (like a whiteboard presentation) and moving a mouse or a control box. Four metaphors are mixed in order to perform this simple interaction with information. Learners have to read these metaphors, and the sum of their combination, in order to make the machine work.

If the physical process of interaction with a computer is so complex, and potentially counter-intuitive (for some learners at least), how much more problematic are subtler aspects of interaction with information when multimedia texts are opened up to learners? For if multimedia is as flexible and open as often suggested, and consists essentially of metaphors for interaction with information, how much more room is there for bad or poor learning?

Of course, designers of learning systems recognise such potential problems, and so limit radically what might have been offered to the learner in order to guide the learner, explicitly and implicitly, on the 'good' choices. For example, a multimedia encyclopaedia typically will allow leaps from a timeline to a historical biography (the implicit model being 'history is made up of the lives of people') but not from 1066 to eye surgery, despite the obvious connection for any child with a morbid imagination. Of course, an on-line medical history, might well do this, but it will ignore the possible link between Harold's death and the mathematics of curves ('how do I fire an arrow that will descend like this?'), the linguistics of names ('what does Senlac mean?') or equestrianism ('why did the Normans ride horses and the Saxons fight on foot?').

In order to cope with the complexity of computer interaction, much of the potential richness of multimedia resources is missed.

Who is interacting with what?

We might also ask who or what the learner is actually interacting with. Deceptively, we see fingers hitting keys and eyes receiving images, and believe this physical interaction is necessarily equivalent to learning interaction. But does turning a page mean that a reader is interacting with an author?

How the learner interacts with any learning materials depends in part on the materials and in part on the educational intent of that interaction. In Seymour Papert's terms (1994), that interaction may be essentially instructionalist (the instructor controls available information and resources to ensure certain lessons, skills or data are learned) or constructivist (learners have freedom to construct meaning for themselves by open interaction in learning environments, building learning out of what they feel valuable in the resources offered).

Multimedia systems appear to offer learners interaction of constructivist kinds: children, we say, can interact with environments and discover things within them. But are the prevalent designs of educational multimedia truly constructivist? Do they permit learners to interact with the information they want in ways which are meaningful to them, or are they essentially instructionalist, guided and restricted interactions, directed to narrow learning goals?

Arguably, a learner interacts not with the computer system, but *through* the computer system (Bench-Capon and McEnery, 1989). The interaction is with the designer of the learning system. The problem of identifying the designer, like the authorial problem in literature, does not invalidate the claim that it is the originator of the learning system, whether individual, team, corporate entity, ideology or culture (i.e. whatever claim we make about authorship) with whom the learner is actually interacting. As authors of text must make guesses about what will amuse, please, inform, educate, puzzle, enthral, excite, stimulate, reward their audiences, so designers of computer systems face exactly the same problems. They are designing for a hypothetical audience, using some form of model of their learners, and of learning, and building a system which encapsulates that model.

In other words, an interactive system is seldom 'an information resource' that learners can explore, sample and interact with in any way they desire. Rather it is a learning model, of greater or lesser complexity, within which learners are expected to acquire certain benefits, no matter what interactions they choose. Their interactions are limited by the designers' model of learning.

Of course, just like any other encyclopaedic information resource, the learning that results from using a multimedia system will be as much a function of how it is used as of what it contains, and that use will to a large extent be determined by the learning context—the task, the guidance given by a tutor, the social rewards of minor success, the links to what is already known and what is pertinent in the classroom. So interaction may be merely a minor benefit of these computer systems, it is the information content which is the real benefit of multimedia systems, not the multimedia interaction.

Which, of all possible interactions within a computer multimedia system is the learner actually able to carry out for any particular system? Here are some test questions:

- can they interact with different versions of the software (e.g. if it is self-updating)?
- can they interact directly with data (e.g. to alter the model of a simulation or to add their own graphics to a gallery)?
- can they interact with other software (e.g. to move information from the multimedia system into their essay, or to import an animation created with a graphics package)?
- can they interact with other users through the system, or alongside the system (e.g. multimedia which offers a group interface, or which contains a 'bulletin board' for users, or which encourages group skills for fuller use)?

- can they interact with an external physical environment in some way (e.g. which connects in flexible ways to different aspects of the curriculum, or which interfaces to other devices such as a synthesiser or a robot arm)?
- can they interact with distributed external information sources (such as direct access to other CD-Roms, connection to the library catalogue, or access to the contents of all the learner's previous schoolwork)?

All these potential interactions are possible through existing technology, but they are determined by a designer or author, who permits, or disallows, such interactions through the system's design.

Very often a so-called 'interactive system' merely consists of relatively routine control operations for the user to perform. Essentially these are static information systems, divided into chunks, whose interactivity merely allows the user to access chunks in particular orders or ways. Such systems are little more functional than a paper-based encyclopaedia or other information source which organise information hierarchically, alphabetically, historically, by topic, indexically, through contents pages and so on. These systems follow an instructional model of education, and are the most prevalent multimedia systems.

However, it is impossible to deny that multimedia offers improved interactivity over such paper systems, largely in three areas:

- speed of access to any chunk of information
- lack of need for user to make detailed searches (the computer already has the links or can make the search on the user's behalf)
- increased number of possible routes through, or organisational access points to, the information.

These are all benefits of computer-based multimedia information systems which, when coupled with good graphics, audio output and moving images make for exciting, entertaining and useful tools, but they still do not offer much in the functions of interaction that paper systems cannot offer. If we expect learning to result from interaction, interactions of this kind are unlikely to teach much more than information retrieval skills.

Multimedia systems will, of course, pass information to the learner, who may acquire and retain it, though we have to be careful if we maintain that learning information from a multimedia system is as good or better than conventional information access. Simply because a system is more readily used by learners and preferred by them, does not mean they will acquire or retain information any better by using it. They may, in fact, spend more time 'playing' with the system than taking any real information from it. Speed of access may be counter-productive; do learners read all the information they access, or do they skip rapidly to other partially-digested topics? Does the learner pursue logical routes through the information-base or do they make random hops as particular tools or quirks of the information momentarily attract?

What might multimedia interaction be?

So is there a different, more engaging type of interaction that multimedia systems might provide and which may have greater benefits in learning? What do we want children to learn from interacting with computers? Is it merely the facts, information, knowledge, truth or beliefs contained in a system (in which case the interaction is something of a sham, merely a test of the learner's ability to discover what is hidden), or do we want learners to develop new skills in the process of this interaction?

Often, ICT skills, are touted as the rationale for using computers in teaching. Without doubt subsets of possible ICT skills will benefit most children in subsequent life. It is also fair to say that, as computers will be embedded in the everyday life of information and transaction, so it is reasonable to embed them in ordinary learning. But surely this is a limited aim, short-sighted in the real environment they may eventually be interacting with?

Firstly the range of skills they will need is unlikely to be acquired in a classroom which does not explicitly address those skills. For example, the notion of getting information out of a database can be learned, but the process of using a particular database (let us say the TAPS database of training opportunities, widely available as a public information resource) requires a specific encounter with that particular interface.

Secondly teachers are unlikely to be able to predict the actual skills they will need with the technology they really face. What, in the last five years' teaching, has prepared young shoppers to scan their own prices at Safeway, to program their videos and TVs for Channel 5, to interrogate BT information systems using voice and tone control, to word process using IBM's speech interface, or to use voicemail? Virtually nothing, because by the time these technologies are widespread in the classroom, and recognised as either reasonable learning tools or objects of learning, they are already being superseded in the world outside, that the students are preparing for.

So, whilst we do need basic ICT skills, awareness of possible technologies, and to break down as much as possible any resistance or phobia (assuming we accept that 'ICT skills are needed for the future world'), to regard any particular set of skills, technologies or interactions as privileged on these grounds is probably a mistake. Unless something else of real educational value goes alongside the ICT skills, such encounters are likely to be of limited value.

We could look to other kinds of learning interaction as a way forward for multimedia education.

Children, of course, may learn more from informal interaction than any amount of formal teaching, the richness of playground lore is one evidence of this; the unexpected, creative leaps they can make from one context into another also belie the belief that any form of *programmed* interaction can match the flexibility of the learner.

Interacting with peers may be a varied and risky business, and is subject to the constraints and rewards of all interpersonal by-play, but it is necessarily one of the prime sources of information for all learners. From 5-year-olds sharing playground rhymes and roller-blade techniques, through adolescents comparing the mythologies of Games Workshop, Dungeons and Dragons and Ancient Greece, to undergraduates sharing notes and revision sessions interspersed with gossip and confessions: all learners participate in groups which may be, for them, the primary source of valid information. What can a multimedia computer offer to match this complex and varied interaction?

The answer, probably, is not much. The machine may, through the amount of data, complexity of programming, intricacy of design and richness of internal interaction between components provide a detailed and varied environment for a child to interact with. But this interaction is still largely about 'receiving the design'.

Certainly exploration and discovery occur in such environments, and they are found both challenging and interesting by children. The avid players of Doom and Discworld show this, though the educational value of both these games may be questioned. Yet there is always a point within such environments where the child reaches a 'so what' level. Yes, they want to complete it, to solve the puzzle and finish the game, and they may get further pleasure subsequently, the pleasure of recognition, by re-doing the puzzle. But before they have reached this point, they have already abandoned 'pure' interaction with a rich and complex environment for something beyond the computer which builds-in other people: 'How far have you got? How did you do this bit? I did it this way. How many did you kill? Is that all? But if you go that way first, you'll find the gremlin. I know that when I do this, such and such happens, but I don't understand why.'

'I don't understand why'. Trading information with other people about such environments is part of their value, possibly their main value. The primary interaction for many players may not be with the game, but around the game. Educationally, we might seek to profit from this. It does not matter where you go, as long as you come back. It does not matter what you do, as long as someone else profits from your doing. If we take this view, then any complex and challenging computer environment could be of educational value, if it permits interactions of the kinds that the learner wants. As Ben Shneiderman advises authors of hypermedia, users, not authors, know what the system should do for them:

> Know the users and their tasks: users are a vital source of ideas and feedback....Study the target population of users carefully to make certain you know how your system will really be used.
>
> Shneiderman, 1992

In a similar vein why not ask what children actually want from computer interaction? What are the pleasures they seek and return to? If the educational aim is to encourage interaction with information which is meaningful to the learner, and to facilitate interactions between children through and around the computer, then the learners' desires for worthwhile interaction are paramount.

Although different children have different desires and predilections, as well as different abilities, younger children who have experienced video games tend to perceive other complex software as games. This is true not merely for software which has obvious features in common with computer games, such as high quality graphics or exploratory features, but also can be true of applications software such as graphics packages and word processors. Computer games set up expectations for interaction which educational multimedia often does not satisfy.

For example, children like not merely to 'explore data' but to interact directly with the environment they find themselves in, whether it is a representation of a concrete reality (such as a room) or a more abstract representation, such as a timeline. They will click on items randomly, move them about, seek to combine, delete, edit, manipulate, zoom, colour and experiment, not merely with the data, or the representation, but with all aspects of the system they are presented with. If nothing happens as the result of experiment, (because the designer had a limited view of 'legitimate interactions') then frustration may result, and certainly boredom is rapidly reached when the learners feel they are compelled to operate only in a single way. No matter how good the graphics and how realistic the sound, a learning system lowers motivation if learners feel they are being told what they can do, rather than allowed to play and learn by using their own strategies.

At the same time, we have to recognise that the strategies of play may themselves be limited. If a child has learned only a limited range of possible interactions with computers, for example, by playing only a limited genre of computer games, then their habits of computer interaction may themselves be restricting (see, for example, discussions by Wegerif and by Grove and Williams elsewhere in this volume). It may be that one of the most valuable functions of multimedia would be to open up the universe to a child, to enable new, unfamiliar interactions, new ways to play with the world.

The dangers of limited metaphor

In a low key way, I would like to suggest that there may be learning drawbacks through the use of limited metaphors of computer interaction, and that learners may develop false notions by virtue of the interactions they engage in with computers, because of the reinforcing message of these metaphors. In particular they may get false ideas of:

- what computers are
- information systems and how they might work
- other forms of interaction, such as interaction with other people.

A computer is not merely an electronic page-turning system that forces you to suspend your activities every so often whilst it compels you to watch an uneditable animation. Nor is it merely a button-pressing device which automatically delivers what you require. Information is not necessarily neatly packaged in discrete chunks that are readily harmonious and all fit a particular idea of reality. Information may be fragmented, problematic, disorganised, contradictory and of doubtful value. It can be hard to find, badly indexed, crudely organised and of marginal relevance. It may need sorting, editing, recombining, changing or testing.

Human interaction, and indeed other human activities, may be described by analogy with computer interaction and processes. People often do this, children in particular. (Williams, 1990). This does not necessarily mean, of course, that they see those interactions and processes as equivalent, imagination serves more subtle purposes than merely literalism. Nevertheless, our prevalent metaphors offer us both enabling and limiting tools, and it may be that the kind of interactions computers permit can shape the predilections of interpersonal interaction. If the interaction with a computer is rich, subtle, rewarding and has an interpersonal flavour, learners might readily develop communication skills which could transfer to social situations. If it is narrow, highly structured, well chunked and follows a predetermined pattern, learners may find acquiring information from real people in the real world something of a shock. Consider the interactions of market researchers in the high street with the general public, and compare this with the model of interaction a computer system offers.

Why is multimedia interaction limited?

A reason sometimes advanced for the limits of human-computer interaction is lack of knowledge. We simply do not know enough about human interaction to be able to make computers do the same, and where we do know about human interaction, we do not know how to make computers act the same way. Such an argument has to be accepted. However, knowledge of interaction is much greater than that implemented in commercial computer systems. The real limits of excitingly interactive education multimedia are those of cost.

Interaction implies choice and choice implies cost.

Every choice that a user might wish to make must be programmed into the system as a possible activity. So, if each step or each activity of the learner is planned in advance, each having many choices, so that all choices are entirely constrained by the tutor/designer, then the cost is great. For every one choice that a user might make, a second has to be implemented for them not to choose. The benefit might not justify the expense as the learners may not make all possible choices (indeed are unlikely to do so) and therefore there will always be elements of such systems that particular learners do not use and which, therefore, from that point of view, use wasted resources.

This means that richly flexible systems can only be justified if there is a large (or alternatively a rich) user base. Much better, therefore, from the developer's point of view, to limit choice in the first place to 'correct', 'meaningful' or 'best' choices.

Choice also, therefore, has costs in development and system resources: the more choices are required, the longer the design time, the more elaborate the programming, and the greater the storage requirement. This means there is a very direct trade-off between sophistication of interaction and cost of a system, which in turn requires that educational technology undergo systematic evaluation to determine if the cleverness of the possible interactions justify the resources expended. Can we identify the learning benefits of particular forms of interaction sufficiently well to cost them? As yet, evaluation of learning technology is not able to do this. In short, we do not know what makes for efficient interaction through learning systems, though we may have some ideas about what kinds of systems are effective, given a typology of learning outcomes.

User choice may also create design problems and, if they are not adequately solved, user problems. Classic problems result from poorly designed hyperspaces, where information is not fully linked, or not properly signposted, creating possible loops in the system, holes in hyperspace (i.e. information which is inaccessible) and users who get lost or confused by the system they are using. If learners are perplexed by the fact that their word processor offers them many different ways to alter text, how much more confused may they be by systems which offer multiple routes into unfamiliar information?

In response to the problem of designing unbounded information resources, and the potential black hole of development costs, both industry and education seek to establish cost-effective mechanisms for development based on re-usable material. Within the Technology in Learning and Teaching Programme, for example, one of the emphases was on portable materials, educational design that could be used in many different contexts. Templates for educational materials are much under discussion amongst developers.

A template is a software system (probably written using multimedia authoring tools such as *Toolbook* or *Macromind Director*) which has no particular application, but into which information can be 'poured' by non-expert developers so that, in principle, any educator could develop multimedia materials with no training in development (Tait, 1995). This, of course, is a powerful and useful idea for hard-pressed educators who cannot afford the time or money for commercial training. I, as a teacher, may not have time to develop the skills to build a learning application from scratch. But if I receive a template today into which I can put educational materials I have readily to hand, and these then become both more motivating and more accessible to my students, I can probably find the time to do this minimal task.

In a similar way, commercial interests look to develop strategies for software development which are re-usable. Ideally they would create methods of multimedia development which can be exactly replicated every time a new package is created. This minimises development time and the commitment of resources, requires less skilled personnel (who have merely to apply someone else's approach rather than develop an original design of their own) and enables all the cost benefits of the production line as opposed to the creative design studio.

However, the template approach to multimedia design, driven by cost effectiveness, necessarily drives in turn a philosophy of limited interaction for the user. A template which can be used for teaching geography, let us say, will only be designed with geographer-like interactions. We might expect a universal geography template to include functions such as zooming in and out on maps, clicking on maps to view statistical information or photographs of locations, supported by textual information and questions/answers on evidence within the maps. Apply this template to the teaching of English.

The more general the template, the less relevant its functional interactions will be to any particular area. The more particular the template, the more relevant will be the user interactions it offers, but the less applicable to other educational areas. A generic template might be so universal that no-one finds it relevant. A particular template may be so particular that no-one can use it in a different context. This tension is now being experienced by those disseminating the outputs from TLTP.

At its cost-effective extreme, commerce is driven to make the widest claims for the simplest models of multimedia interaction. A recent publication (Gayeski, 1995) offers a 'hands-on workbook' which 'contains strategies, models, examples, checklists and worksheets that guide you through the process of analysing needs, selecting strategies and formats, budgeting, scripting and flowcharting, authoring, and managing computer-based information and instruction'. This sounds wonderful. But when we examine the heart of the text, that on actual implementation of interaction, we find it is reduced to flowcharts, made up of information, questions and feedback. Instructional design is precisely that – instruction. Interaction is no more than the choice of which questions to receive when, and whether to give the right or wrong answers.

The consequence of the imperative of cost is that templates tend to seek models of the lowest common denominator for interaction. What do people want to do, whatever subject they are being taught? Well, they go forward in the materials, they go back. They may need to return to the start. They use an index, they need summaries, they need help. They might need remedial pages at structured points. They may need to mark 'pages' they have already visited. They may need a glossary of difficult terms which they can call up when they like. They may like to call up illustrative pictures of particular text.

Does this sound like any system you know?

This is not to deride such systems. They can be as useful as, perhaps more so than, the paper-based resources they are based on. I have built such systems and found such usefulness. (Williams, 1992). But it would be wrong to pretend that these are excitingly interactive, offering all sorts of new and different kinds of educational experiences or enriching the learner's experience widely beyond what could be acquired from other sources. A library and a video recorder can offer very much the same educational value, if the learner has the freedom to use them.

A metaphor for multimedia interaction may itself be almost as restrictive as a template, though perhaps more subtly. *Toolbook* is widely used on PCs for in-house development of training and educational materials. *Toolbook* is based on the metaphor of the book (others, such as *Hypercard* on a stack of reference cards, and *Director* on a theatre). Such a metaphor is necessary to enable new users to get to grips with unfamiliar concepts. (If we were told we were in the business of placing semantic icons in an unbounded cyberspace, we might find it more difficult to grasp). But the consequence of the metaphor is that the authoring system tends to limit its horizons, and the developers tend to go along with the model offered, even when excursions beyond the metaphor are possible.

So the typical *Toolbook* system consists of pages, through which users move forwards or back, sometimes via a contents page of some kind. Occasionally they move laterally through hyper-links. But the lateral jump is not the structure of the system. The book is the structure, and the hypertext of hyperlinks is an addition, an extra kind of interaction with the information, certainly, but also a secondary one. The hyperlinks are there between pages that already have their structure, books with links are made by *Toolbook* programmers, not hyper-documents.

Beyond passive interaction

Even if *Toolbook* applications were to take full advantage of their medium, they would still provide largely passive links for users to follow. The foundation metaphor of the book is precisely this. Although authors like Bob Cotton and Richard Oliver (Cotton & Oliver, 1993) may maintain that 'the content of a book viewed on computer monitor or television screen is not a book', it is not interactive multimedia either.

Where is the authoring system which encourages developers to provide learning systems which the users build for themselves, for example? We have the work of Papert and our own experience that tell us learning is doing, learning is experiment, learning is discovery, learning is interaction with a problematised environment. You learn how to build Lego houses by building Lego houses. (And,

I might add, you learn how to build multimedia systems by building multimedia systems.) Which multimedia educational resources allow the learner to build their own routes through the materials, to delete or over-ride those created by the designer? Which allow the users to add their own comments, modifications and original material? Which allow groups of users to build, or reconstruct the information resource collectively? Which offer visual and aural resources which can be manipulated and linked in ways which the learners find interesting?

I do not know of many. My two answers, both offered with hesitation, would be the Internet and *Klik and Play* (Europress, 1995).

To take information off the WWW is not difficult, given the hardware and the software. But to place information on the Web, and to link and adapt it to your own satisfaction requires additional systems and training in their use (the complexity of the training depending on the design needs and the actual system in use). Now, HTML (the code generally used to write World Wide Web documents) is no more difficult to learn or use than any other language or code. The recent availability of authoring tools which create HTML script make the task rather more friendly. Nevertheless, it is much easier for me (as a user and learner) to receive other people's information than it is for me to interact with that information, to modify, develop, and add to it. Why should this be?

The reason, perhaps, is that the Web is driven by information providers, in the widest sense (from self-advertisers to subversives) who are a technical elite. They have a model of information which is essentially still the traditional authorial model of me, the author, telling you, the reader, what the truth is. You can receive my information, and you can tell other people about it ('please publish my address') but you cannot do anything with it, because then I would not own it, it would not be mine, I would no longer be author and lose my authority.

The classroom teacher is, of course, the author of learning with authority over information, in precisely this traditional way. Whilst we may be willing to step back a little, to relax the authority, we are driven by a syllabus, a National Curriculum or by our own self-interest as professional educators with privileged expertise not to relinquish it. Whilst multimedia technologies might offer us the chance to abandon self-interest and curricula to the interests of the learner, we are unlikely to make much use of that chance.

For what we, as professional educators, see is the learner merely engaged in clicking and playing. You are probably not familiar with the *Klik and Play* software system, which is not usually touted as an educational tool. It is a game-building system for children. Actually rather limited, in terms of the underlying game structures that the child can adopt (and thus another example of a potentially open-ended system limited by its own metaphors), it nevertheless offers creative children hours of fun designing icons, pasting existing icons, attaching sounds to images, assigning pathways to icons across screens and determining the conditions under which certain events occur (collision detection, for example). They can import external files (sounds, images) as well as adapt the existing library of resources. Apart from the obvious learning payoff of using *Klik and Play* (the child learns more about how computer games, and by extension computers, work), what is the educational benefit of such a system?

None, if what you want is the transfer of particular information in particular, organised chunks into the head of learner. Lots, if what you want to see is creativity, imagination, links and leaps between fragments of meaning beyond those the teacher might conceive. We might not understand creativity, but we know it when we see it. If, like Koestler, we can accept that the Act of Creation (Koestler, 1966), whether humour, science or art is essentially the synthesis of two previously unlinked terms, then *Klik and Play* offers tools for creative learning. Any sound may be attached to any object. Any object may have any form. Any object may have any behaviour. Any two objects may interact in any way. The child creates a world and watches it happen. And, if they do not like what happens, they remake the world. If they do not like the sounds, they can sample their own. If they do not like the images, they redraw them, or steal them from elsewhere.

This tool is sufficient to allow the child to replicate a reality, so the user can settle for simulation. It could also be used as a communication device between learners, or learners and tutors, for the child could use it as any other representation or modelling device. For example, the child might be asked to model their living room and what goes on there; or to model the movement of the planets in the solar system. Their models become means of presenting their understanding and ideas for others to compare.

I have not seen this tool used in this way. However, I suspect that if it was, we would see strikingly original conceptions of how things work and what things are. Instead of multimedia which is directed to telling the child what to think, albeit in attractive and motivating ways, we have multimedia which allows the child to express what they already think, and to create new ideas and present them to the world. This is achieved by allowing the child to interact directly with the system and its data, to become an author, to assemble their own data, and to take nothing in the computer system for granted.

Of course, I am overstating my case here. *Klik and Play* is actually a somewhat limited two dimensional game creation system, creating games of limited enjoyment. But that is not the point. The child does not get the main pleasure from playing the game, but from conceiving the game and making it happen. The child is engaged in the process, not the product. Yet there is always a product, a demonstrable and, if you like, assessable, end point.

Klik and Play also offers no predetermined textbase. Its information is entirely visual and aural, and it is 'comic game' information, rather than anything we might associate with legitimate learning. If a child wants to develop something deeper in this respect, they presumably have to acquire the authoring skills necessary to work with *Toolbook* or *Director*. (Parenthetically let me say that my 10-year-old son acquired all the *Toolbook* skills necessary to create a five page interactive birthday card, complete with animation, sound and humour, in a day, which was rather less time than it took me to acquire those same skills.)

Klik and Play perhaps gives some hints of how the WWW should look (and be used) for children to take control of information and play with it, to learn by interacting with the burgeoning multimedia resource. Authoring tools offer some simplification, and some new possibilities to learners, but they fall short of what might be offered, and are fundamentally limited by the pervasive metaphor of the document.

If you visit a Website, what you typically see is a graphic hook of some kind, followed by a long, scrolling page interspersed with buttons (to other pages or sites) and graphics. Occasionally there will be other resource files which must either be downloaded or may run when activated, if the reader's computer runs appropriate software. The Web is really a big library of long, illustrated documents, each document referenced to others, and containing images and clips that you receive and consume. The new authoring tools let you write such documents, marking up your text into HTML, as if you were using a sophisticated word processor or Desktop Publishing system. The concept of this process is at least nineteenth century: the author writes, someone else (the computer in this case) marks it up for publication.

We can see why this is the dominant metaphor, but why should it continue to be? If the Web is a web of information, why cannot that information be presented and used in any way we might want to manipulate such information? If it is a sound file resource, then why is the metaphor not that of a recording studio? If the Web is a network of video clips, why are authoring tools not essentially film sets? Why is the metaphor not one of stage or screen or animation studio? Better yet, why is it not all of these?

How many video clips on the Web branch within themselves to other clips? How many sound files can be mixed or edited as they run? How many animations or images on other websites can you add to or amend at their host site? How many interactive conferencing systems run without a central manager? Why can't a learner click on a resource and play with it, then bend, staple, fold and perhaps mutilate it into their own point and click document? Why can't a learner wander the Web and compile, not a history of where they went (like a card index of the books I've read) but, automatically, a new document made up of everything encountered? Why are users restricted to models of old technologies, when what they are using is a technology previously unconceived?

There are technical answers to these questions, but they are apologies rather than reasons – the technical issues can all be addressed. The true answers to the questions perhaps hint why multimedia systems offer typically passive interaction ('here are ten different ways you can receive our information') rather than active engagement with the data or the program. If you offer the learner these possibilities, you offer control, you offer authority. You allow the child to 'mess about', to 'make mistakes', to 'waste their time', to 'damage the system' because these are activities which (a) interest and engage them and (b) which necessarily make the modified system different from the author's intent.

Conclusion

Hypermedia allows everything to be linked to everything, just like human imagination. But teaching (as opposed to learning) requires rather narrower perspectives. Whilst the information within educational multimedia is extensive, attractive and complex, the interactions with and within that information, allowed by the authors of such systems, fail to realise the educational potential of the medium. We need more radical reviews of what learning might be through multimedia, combined with innovation in design (Stringer, 1997), to realise more fully the true potential for educational interaction in multimedia.

References

Bench-Capon, T. J. M. & McEnery, A. M. 'People interact through computers not with them', *Interacting with Computers* **1** (1) (1989), April: 31-38.

Europress. *Klik and Play*, 1995.

Gayeski, Diane M. *Designing Multimedia: An Interactive Toolkit.* Future Systems, 1995.

Koestler, Arthur *The Act of Creation*. Pan Books, 1966.

Papert, Seymour *The Children's Machine: Rethinking School in the Age of the Computer*. Harvester Wheatsheaf, 1994.

Reynolds, Angus & Iwinski, Thomas *Multimedia Training: Developing Technology-Based Systems*. McGraw Hill, 1996.

Shneiderman, Ben, 'Authoring and editing hypertext.' In Hartley, James (ed) *Technology and Writing*, (1992) 162-8. Jessica Kingsley Publishers.

Stringer, Roy, Fruit Gum Guide to the Hypermedia Universe. *The Times* May 9 (1997): 36.

Tait, Kenneth, 'Are Templates What We Need? How to Allow Courseware to Evolve.' In *Proceedings of 1994 Toolbook User Conference*, (1995) 91-100. Centre for Computing in Economics, University of Bristol, UK.

Williams, Noel 'Computerspeak: the Language of New Technology.' In Williams, Noel & Hartley, Peter (eds) *Technology in Human Communication*, (1990) 135-145. Pinter Publishers.

Williams, Noel 'A Hypertext Open Learning System for Writers', *Instructional Science* **21** (1992): 125-138.

Explorations in Virtual History

Jonathan Grove and Noel Williams

Virtual realities

'Virtual reality' is a widely used term, but it remains ill-defined. An unfortunate necessity of working in this area is that the medium itself is the constant subject of re-definition. 'Virtual reality' is not really a single phenomenon but an umbrella term, a catchall, or what Brenda Laurel (1993) refers to as a 'monolithic icon for a complex network of ideas'. In the range of literature on the subject, virtual reality VR) is described as a tool, a concept, an experience, various combinations of chnologies and a psychological variable. This is not helpful. For our purposes, define VR as an interactive three-dimensional representation of a real or ract space that is displayed by a computer. This is not *the* definition, but *our* king definition. Other researchers certainly describe VR in different terms.

Many researchers argue that VR technology is a potentially powerful cational tool. In her paper 'Virtual reality: potentials and challenges' Meredith ken (1991) describes this potential as encapsulated in five features. To aphrase, these are:

- *VR* is experiential: learners using VR 'feel a sense of presence within a virtual world'
- *VR* allows natural interaction with information: the skills of the VR world are the skills of the real world
- *VR* is a shared experience: virtual worlds can provide social contexts
- *VR* allows unique capabilities: the power of VR allows control of 'time, scale and physics'
- *VR* can be tailored to individuals: information can be represented differently for different learners.

Individually each of these features could make for exciting educational software. Together they may represent a significant innovation in educational technology. However, the operative word here is 'may'. Initially most literature on VR discussed its potential in very speculative terms. Only in the last couple of years have researchers really begun to address pragmatic and theoretical questions to determine the real value of VR technologies. Now researchers such as Winn (1992), Osberg (1992), Byrne (1993), Pantelidis (1993) and McClellan (1991), and well-funded centres such as the Human Interface Technology Lab at Washington University, are all considering how the educational potential of virtual reality might be best understood and utilised.

VR clearly does have great potential as an educational tool. However, that potential is dependent on educators being able to locate the technology in a sound educational framework, a framework which takes into account both the pedagogic and pragmatic aspects of implementing VR in the classroom.

The work at Sheffield Hallam University aims to contribute to such a paradigm, particularly for the teaching of History, as history seems a clear area that might benefit from the ability to explore virtual spaces. Pluckrose argues:

> In my experience young children are able to identify with the past more readily if they are taken to places which have some historical significance, rather than if their exposure to history is limited to the page of a book. Imagination and a sensitivity to the human condition within a framework of time past is more easily fostered in young children if their senses can be sharpened. The priests' hole in a book can never be as threateningly claustrophobic as it was in reality for Ralph at Oxburgh Hall.
>
> (1994)

Whilst Pluckrose does not have technology in mind here, VR obviously fits i with his thinking. VR allows the learner to be 'taken to places', modern or ancie without them leaving the classroom. Such experience cannot be supplied by historical multimedia packages which only allow limited interaction with information presented in a two-dimensional format. Multimedia brings 'information', but VR offers 'exploration'.

Our main area of interest is in encouraging children to use VR in small group We believe that one of VR's principal strengths is the way in which it can be used to stimulate talk between learners. The fundamental importance of talk for learning has been highlighted by many researchers. In the UK the Russian psycho linguist Vygotsky (1978) has been highly influential in the development of linguistic accounts of the relationship between talk and learning. In the USA, these ideas have been absorbed into broader frameworks for teaching and learning, many of which would come under the umbrella of 'constructivism' (a topic we return to at the end of this paper).

The focus of our work then is 'learning talk', and one of the main purposes of the case study was to promote dialogue (primarily, but not exclusively, about history) between learners.

In the following section we look at some of the issues that came to light during our case study. As suggested earlier these are not finalised results. Rather they are initial observations which we think are of interest.

The Case Study: some preliminary results

Concretely, the focus of our research is a historical virtual environment. It is a 3D model of a large ancient Greek residence, built using Superscape plc's virtual reality design software VRT. This is a tool for constructing virtual environments that runs on a standard desktop multimedia PC. The tool is expensive and difficult to learn, yet quite complex environments can be built with limited interaction. Users can view a static or moving object or group of objects from a variety of viewpoints. They can open doors and windows, work artefacts in various ways and hear sounds associated with those objects. The system also allows the inclusion of rudimentary multimedia elements such as text boxes, hyperlinks and photographs. The environment itself is designed to locate with part of the National Curriculum Key Stage 2 Unit 'The Ancient Greeks'.

The building in the model is two floors high, with nine rooms and a courtyard. The rooms contain numerous objects, many of which are modelled on photographs of archaeological finds. Learners can explore the building, investigate rooms and objects and can of course fit those interactions in a context defined by a teacher, seeking to answer questions or discover information which has been represented n the environment. However, it is a non-immersive environment. That is to say, it a three-dimensional model presented on a two-dimensional screen, controlled rough mouse and keyboard. We made no use of datagloves, stereoscopic dsets or other immersive technologies.

This VR model was used as a virtual history lesson with a class of thirty 8-10-ar-old learners, of mixed sex and ability. The children were given training in up work, in the use of Exploratory Talk (ET) to address tasks, and were also en an introduction to the VR interface, using the example of a small virtual fice. (For an explanation of Exploratory Talk see Wegerif in the present volume). hey then worked in groups of three, exploring the Greek House environment and ddressing specific tasks which were set for them. They were encouraged to see he environment as a collection of historical evidence, not merely to wander imlessly, and to treat that evidence as a source of speculation about the life of the ncient Greeks. After their experience of VR they were asked to complete a questionnaire on their experiences and to write a descriptive report. Their interactions were also video-taped during use of the system.

Most of the learners in the case study enjoyed their experience of VR. This is highlighted in the questionnaires and reports they completed after they had explored the environment. For example when asked if they liked exploring the Greek House, 28 learners ticked 'a lot' and the remaining two ticked 'quite a lot' (using a five point Lickert scale, rated from 'a lot' to 'I hated it').

All but one child said they would prefer learning using VR to studying from a book. Twenty children claimed that they would prefer using VR to a 'regular' computer.

These findings are reinforced by comments in written reports produced by the children. They were encouraged to write down information such as what they did during the VR sessions, what they liked or disliked about VR and what went wrong. Three indicative quotes are:

> The VR project was better than I expected, but next time I would like to explore a Victorian House
>
> It was so much fun exploring the rooms and looking at pots and furniture
>
> I answered the questions because I enjoyed the VR tasks so much. If I was going on the VR again I would like to go into a palace to see what it was like

Other ideas for virtual environments included an Egyptian pyramid and a cave.

Of course the principal aim of school is not to offer enjoyable experiences but to encourage learning, and the case study produced plenty of evidence that VR can offer rewarding learning experiences as well. Firstly this is apparent in the talk that took place between group members while they used the virtual environment. A wide range of topics was discussed including the contents of the rooms, their purpose, the layout of the house and the human viewpoint's position in relation to the crude map they were supplied with. Groups also discussed aspects of spelling and grammar when completing the written parts of their tasks. Some learners expressed an interest in learning how the house was constructed, and one or two subjects demonstrated considerable understanding of the world building process

The questionnaire responses also suggest that learning took place. For example, in response to the question 'Do you think you learnt anything?' two thirds of the subjects said that they had. When asked what they thought they had learnt, answers ranged from 'how to use a mouse' to 'I learnt that the Greeks lived not like us' and 'about what the beds and everything looked like'.

Perhaps more revealing are the written reports which gave the subjects the chance to write unrestrictedly about their experience. Among other things these pieces of writing contain references to the content of the environment. They are a testimony to the immediacy of VR as a learning tool. This is illustrated by the following quotes.

> I opened the door and it was a Greek carpet. We thought it was telling us that the Greeks used to knit and fight because there was a picture of Greek soldiers.
>
> It had a box in the corner. I clicked on the box and the box opened. A flute came out of the box and went back in the box. We went into another room with furniture in it. And we went into another room with furniture in it. And we went into another room with wine in the room and pots and pans.

The children also decorated their reports with illustrations and some of these are quite revealing. For example, one learner fell off the virtual house's second-floor balcony (though we are not sure how she did it). She remembered this, referred to it in her report and drew a computer screen displaying a virtual balcony.

From a linguistic perspective, one aspect of the study being explored is the use of 'indexical' terms among the learners. In essence indexical terms are linguistic items which are wholly dependent on context for meaning. The words 'I', 'there' 'here', 'tomorrow', 'yesterday' are all indexical. Review of the video data collected from the case study seemed to show frequent use of terms of this type. Indexical terms are a fundamental feature of discourse between people involved in a shared task, and these kinds of terms are quite common when learners use multimedia. For example, utterances such as 'click on that', 'point it over there' are often used.

However we would argue that the use of these terms in the context of multimedia is significantly different to their use in the shared context of VR. In the former, indexical terms have little impact on learning, they are essentially 'external' to the task in hand, being concerned primarily with the management of the group and the machine. But in the latter, utterances such as 'go up there' are fundamental to the learning experience, as 'there' indicates a position within a (conceptualised) three-dimensional space, rather than merely a place on a screen to put the pointer on. This possible distinction needs further exploration.

The particular focus of the study is to examine the discourse of groups of learners who use VR to see the extent to which VR can promote discourse of educational value. There is no doubt that groups interact around their use of VR, but to what extent are those interactions likely to pay off in educational terms? Are they, for example, primarily debates about who should take them where in their travels, or are they directed more towards the educational objectives of the exercise (e.g. coming to conclusions about the historical evidence being offered by the simulation)? We use the concept of Exploratory Talk discussed by Wegerif in a previous chapter to examine this issue.

As the aim of the Greek House case study is to offer 'historical evidence' which can be explored, interpreted and debated by learners, we would hope the groups' discourse showed use of ET. Evidence of ET within their discourse would be evidence that a rich and complex primary learning environment, such as our VR house, provides some stimulus for discovery in learning, for the exploration of concepts and problems through group interaction and for establishing reasoned, consensual judgements about that evidence. In other words, we are not looking to see if the learners necessarily form the correct judgements about the evidence offered to them, but that they treat that evidence as historically valid and debate it in a reasoned way before forming conclusions about it.

Such evidence of ET would suggest that the simulation is seen as 'valid evidence' and that the experience has sufficient reality for the children to want to argue for their own interpretation of what they have seen and 'been to'. We would be prepared to argue that such evidence would show a rich learning environment has been created, an environment which stimulates thought, debate and consideration of historical reality. We are more interested in showing that processes of learning are taking place as a result of using the VR environment, than assessing the historical 'facts' that might have been acquired by such an environment. At the same time, of course, we want our environment to be as accurate and detailed a picture of the historical artefacts as possible. There is less point in debating poor evidence than good evidence.

As yet we have only preliminary findings, but they suggest a wide range in use of ET between the different user groups. Clearly in some instances the environment does prompt the exploration in dialogue we hoped to see. However, it does not do this consistently, and there are clearly cases where children are attempting to use ET more because they had been coached in its use (as part of their initial guidance on group work), rather than because it enabled them to arrive at some consensual judgement on the evidence before them. According to our initial examination of the data we could not suggest that this VR case study, as an exploratory learning environment, necessarily promotes ET for all learners, but it does stimulate some.

However, the data has yet to be subjected to detailed analysis. Given that all learners seem to have enjoyed the experience and many do seem to have learned something, it may be that the variation in use of ET across groups is due to the variation in ability of members of particular groups, or to particular group dynamics.

We also encountered several unforeseen problems, and these may also have complicated the learners' interpretation of the evidence the environment offers. For example, the learners' reaction to the iconography of the environment was unexpected. The design of the virtual environment and case study had not fully taken account of the prior experience of learners. When young learners use a virtual environment for the first time, their understanding of it will be based on their past experience and knowledge of IT and VR. The source of this model is likely to be a mixed bag of TV programmes, magazine articles and their experience of 'VR-type' games such as Doom. These influences clearly conditioned their reaction to the environment.

For example, during the initial training of one class member, using the virtual 'office', she cryptically asked me how she died! There was no apparent threat in the environment she was exploring and yet she assumed that the logic of a computer game was operating. A similar phenomenon was apparent when one group explored the Ancient Greek residence. One room had a stripped wood floor and, to make the environment more realistic, some panels were made lighter than others. The two male members of the group saw this as significant and spent a long time 'clicking' on one of the panels believing it to be a switch of some kind, a mistake they repeated when they noticed a discoloured slab in the courtyard. Now this is not necessarily a problem. However, it clearly distracted them and it suggests that they saw the environment as more game than reality. We need to ask whether encouraging the children to see the experience as a game is beneficial to learning, especially where the ostensible aim of the VR encounter is for the learners to 'experience' a 'reality'.

Another difficulty experienced by the learners relates to the role of object dynamics and properties in the virtual environment. We considered it important that while the learner was viewing the world from a human viewpoint their persona should not be able to do things a real human could not. In particular, users should be forced to walk around objects and not allowed to walk through walls. Such limits contribute to the environment's realism.

Subsequently we judged this as a mistake for two reasons. Firstly as the environment is quite clearly not real and learners do not perceive it as real, such restrictions contribute little, and may be perceived as 'unnecessary'. Secondly object collision will sometimes confuse the learners because it can take time to determine which part of the virtual persona is colliding with which object. In reality if we bump into an object we will be able to feel and hear the collision. Supplementary information can be supplied by quickly glancing around. In VR, supplying feedback of this complexity is very difficult if not impossible and looking around (in non-immersive VR) is not particularly quick. These issues combined with the cognitive load of learning a new interface contributed to a lot of confusion and frustration in some learners.

This will not be a problem if they persevere, but it may put some off using the system altogether. It also means that their discourse tends to be focused on these interface problems, rather than on the historical task before them. It may be that, with the current state of non-immersive VR, the presentation actually places barriers between a learner and the simulation which interfere with learning, rather than facilitating it.

The above observations raise questions around the often-quoted notion that VR is somehow more intuitive to use because it reflects the way in which we interact with the real world. It is possible that until immersive technology really does make the interface 'disappear', then VR may simply replace one set of interface problems with another, more complex set. Assuming that VR is automatically useful because it looks more like reality, may be misleading. The assumption is that because VR is like reality then the learner will instinctively know how to interact with the computer in a meaningful way. However, in its present guise VR (both immersive and desktop) is only a little like reality. It can only represent some of the rules of the world. Its representation of reality is partial, and the learner may have no sense of what is and what is not possible in the virtual environment. There is no framework on which to base their assumptions.

Gibson's (1977) notion of 'affordances' is useful here. According to Gibson, when we perceive our environment we are not concerned with its qualities but what its various elements 'afford' us as human beings. In other words whether an object allows us to walk on it, swim in it, pick it up, eat it etc. The affordances offered by an environment fit with the properties of our own bodies, for example we see something as 'walkable' in relation to our method of locomotion (our feet). This concern, he argues begins early in life.

> Phenomenal objects are not built up of qualities. It is quite the other way around. Objects, more exactly the affordances of objects, are what the infant begins by noticing. The meanings are observed before the substances and surfaces are. Affordances are invariant combinations of variables. And it is only reasonable to suppose that it is easier to perceive an invariant combination than it is to perceive all the variables separately.
>
> (1977)

In these terms the user of a virtual environment has no sense of what the environment affords them, furthermore they have no clear idea of whether the affordances they perceive are reliable interpretations. Another implication of the lack of intuitive understanding of the VR may be that there is no common framework for negotiation between learners. Since if we assume that the world is meaningful, and that the meaningfulness is established through social negotiation based on our similar experience of the world (as constructivists do), then a problematic environment such as VR will remove the baseline not only for a single learner's interpretation, but also for negotiation of its meaning between learners.

Arguably, therefore, VR is a medium that needs to be learned just like any other communicative medium. It needs to be 'read', and users need to be literate in its codes. This argument is advanced by Sherman and Craig (1995), who suggest that users of VR need to learn 'the symbols of the medium' before they can understand its content, a task made difficult by the youth of the medium with its developing and uncertain language.

Another system design issue relates to our inclusion of sound in the environment. For example, we included the chirping of crickets and the sound of a fountain because we again thought it would contribute to the environment's realism. However, it did not have the effect we anticipated. From speaking to the children we found that few noticed the sound and fewer recognised it. They did not link the sounds with the objects on screen. On reflection it seemed obvious that, as there are no virtual crickets in the environment and the water in the fountain did not move (due to technological limitations) there was nothing to link the sounds with their visual sources. The sounds alone, therefore, without a visual clue, were not meaningful to the learners.

Several issues also arose around the way the environment was incorporated into the classroom. Whilst these were specific problems for our particular study, they are nevertheless indicative of the kind of issue that will need to be addressed by any teacher seeking to embed VR in a curriculum. For example, our attempt to teach the children how to work in groups had only a limited impact, primarily we think because the culture of the classroom promoted individualism. Most learners said they would prefer to work in pairs, rather than groups of three. Furthermore, restricting the children to a one hour group session on the system (a purely pragmatic decision), was too limiting. Any future work would need to use the environment in a more open way. Yet these are inevitable problems with such an expensive and intensive learning technology. To justify VR, and to make its potential benefits available in the classroom, learners will have to use it meaningfully in groups and probably for comparatively short periods.

A Proposal for Future Work

Future work with VR needs to explore the possible fit between constructivism and virtual history, and this is where we expect our work to go. The links between constructivism and VR have been discussed by a number of researchers including Bricken (1992) and McLellan (1991). Indeed Winn (1993) has gone so far as to suggest that without the ongoing development of the constructivist paradigm 'the chances are that VR would be little more than another educational gimmick'.

Constructivism is a broad characterisation of an approach founded on two assumptions that knowledge is constructed through social negotiation, and that reality is partly subjective (we all experience the same world but interpret it on the basis of our own knowledge and beliefs). Simply put, constructivists argue for a learner-focused environment, in which learners can explore a knowledge domain and construct knowledge of that domain through a combination of collaboration, discussions with their teacher, self-assessment and reflection. The more traditional didactic approach is regarded as of equal or lesser value than constructivist approaches.

Through a constructivist approach, we hope to make best use of the existing knowledge learners have of VR (even if distorted by popular representations), to use the game-orientation of some learners positively (rather than allowing it to interfere with learning) and to place a VR simulation within the developing dialogue of classroom learning, rather than merely expect it to deliver learning by virtue of its own rich content.

The constructivist paradigm is relatively new, and many issues must be resolved before it can become a sound educational framework. In particular there are many questions about how we can 'design' educational experiences that fit with the constructivist epistemology. Jonassen (1994), among others, has outlined a set of principles to help with this process and we will conclude with a brief summary of how virtual history could perhaps locate with the goals of constructivism.

Jonassen - The principles of constructivist learning environments

Jonassen lists seven features of constructivist learning environments, outlined below. The examples illustrating each of his listed features are ours, and suggest possible lines of development for virtual history projects. Jonassen says that:

> Learning environments should provide multiple versions of reality, thereby representing the natural complexity of the world.

The 'reality' of history is that historians have various ideas about what the past was like. Their representations of the past are an interpretation based on available evidence. Rather than presenting a single version of historical 'fact', VR could be used to represent various versions of history to the learner in order to highlight its interpreted nature.

The simulation of an Ancient Greek residence could be presented to show the relationship between the ruins and what historians believe the site used to look like. Ambiguities in historical interpretation could be highlighted by allowing the learner to switch between different versions of the site. If the virtual representation contains artefacts, the sources for those representations (a combination of actual finds and literary and artistic sources of the period) could highlight the interpreted nature of history.

By doing this we give the learner multiple perspectives on the past, rather than misleading them into believing history is somehow objective. At the same time, there is no pretence that the learner is 'walking' through a 'real' environment and somehow interacting with versions of real objects, so where it is appropriate to do so, the constraints of external physics can be removed, simplifying the interface and the task of learning how to learn with VR.

Learning environments should focus on knowledge construction rather than reproduction

A VR system could present the learner with virtual representations of historical finds, the user must then speculate on their origins, their use, material and method of construction. So learners could be presented with the bare bones of historical research, and confronted with the task of the archaeologist or historian. Instead of being shown other historians' views of a site, they can themselves be encouraged to speculate on how that site might look, based on its ruins and then see how their interpretation relates to that of other historians. In such a use they might even be given a virtual environment that allows them to construct their own version of the simulation (for example, by manipulating virtual building blocks).

Learning environments should present authentic tasks

The interpretation of a historical find and determining the best method of excavating a site are the kinds of tasks that real historians are required to undertake. The task of a VR simulation would be to represent the historians' task, and the tools and processes they use, rather than the outcomes of the task.

Learning environments should provide real world, case-based learning

The VR could be grounded in the real world by linking the package with a real archaeological site and, of course, to other information sources (archives, museums, libraries, on-line historians, historical websites).

Learning environments should focus on reflective practice

Reflective practice involves encouraging learners to appraise their own cognitive models by using a variety of methods. One way of doing this would be to use the computer to record the learner's method of trying to solve a real world historical problem, such as dating an artefact or mapping and excavating a site. This process of problem-solving could be supported by the computer, through the use of various media. Once learners have formulated a solution to a historical problem, they could then compare their approach with that used by a typical expert (this could also be stored in the computer system, along with a record of the learner's approach). The learner could then appraise their mental model in relation to the one used by an expert.

Within a VR environment, this might be addressed by simulating the data, objects, tools and resources a historian might have available to address a particular historical problem, and recording the trace of the learners' interactions with those objects. The constraints imposed on such a system would no longer be those of trying to simulate a real 3D world, but of simulating the process of historical investigation. Here might be an opportunity to combine the video game habits of some learners with the investigative imagination required of historians.

Learning environments should facilitate context and content dependent knowledge construction

VR tasks need to be focused on the application of knowledge and problem-solving methods to real world situations. The emphasis is on the active construction of knowledge, in order to serve a particular purpose. Knowledge is treated as a tool, rather as an abstract domain which is studied for study's sake.

Learning environments should support collaborative construction of knowledge, rather than encourage competition among learners for recognition

Collaboration is an important part of learning, and a necessary part where learning resources have necessarily to be shared, as is likely to be the case with most VR in the classroom. But it is not enough simply to put learners together and expect them to produce the kinds of dialogue which will facilitate problem-solving and knowledge construction. As the early data from our study seems to show, even with prior training on group work and Exploratory Talk, learners in some groups find it difficult to share their knowledge, or to value the knowledge of others.

The productivity of peer learning discourse could probably be enhanced by structuring it in some way, perhaps through the VR simulation. A similar point is made by Dalton (in Rose, 1995) who discovered that the quality of the interactions in a collaborative context is a determining factor in learning. According to Rose, Dalton found that the structuring of learner interactions aided the processes of encoding, decoding and high-level elaboration (where students explain the contents out loud). Rose asserts:

> These studies, suggest that VR technology which fosters collaboration will yield even greater educational benefits. The question for research then becomes: how to encourage meaningful collaboration both inside and outside virtual space? Attention must also be given to how to train instructors to promote desirable interactions when using VR. (p13)

Future work on virtual history would, we think, benefit from incorporating Jonassen's ideas. They provide a more open paradigm for the successful use of VR in the classroom than simply giving a group an environment to wander in.

Conclusion

In this paper we have outlined some of the ideas and preliminary findings that have been generated by our recent case study of a virtual environment for history teaching. We have suggested that a virtual history environment might fit with constructivist views of education, and that such an environment might promote Exploratory Talk, which would be one feature of such educational experience. However, we have also suggested that the current state of non-immersive VR, the perceptions of learners and the difficulty of embedding VR technology in existing curricula and classrooms might at least complicate the learning application of VR and possibly reduce it from a technology of great potential in principle, to one of limited value in practice. Clearly VR offers a richness beyond Multimedia or indeed any other educational technology, but its very complexity may itself interfere with the learning process, resulting in too great a focus on the vehicle of learning and too little on the learning journey.

References

Bricken, M. Virtual Reality Learning Environments: Potentials and Challenges. *Computer Graphics* **25** (3) (1991).

Bricken, W. & Winn, W. 'Designing Virtual Worlds for Use in Mathematics Education: The Example of Experiential Algebra', *Education Technology* **32** (1992).

Byrne, C. *Virtual Reality and Education*. Human Interface Technology Lab report no: R-93-6 available from http://www.hitl.washington.edu/publications/tech-reports, 1993.

Gibson, J. J. *The Ecological Theory of Perception*. Lawrence Erlbaum, 1977.

Grove, J. 'Virtual History: An Outline of Research Being Undertaken at Sheffield Hallam University', *History and Computing* **8** (1) (1996).

Jonassen, D. H. Thinking Technology: Toward a Constructivist Design Model. *Educational Technology* **34** (4) (1994).

Laurel, B. *Computers as Theatre*. Addison Wesley, 1993.

McClellan, H. 'Virtual Environment and Situated Learning', *Multimedia Review*, Autumn 1991.

Mercer, N. 'The Quality of Children's Joint Activity at the Computer', *Journal of Computer Assisted Learning*. **10** (1990).

Osberg, K. M. *Virtual Reality and Education: A Look at Both Sides of the Sword.* Available from the Human Interface Technology Laboratory: Washington D.C., 1992.

Pantelidis, V. S. 'Virtual Reality in the Classroom,' *Educational Technology*, April 1993.

Rose, H. *Assessing Learning in VR: Towards Developing a Paradigm Virtual Reality Roving Vehicles Project (VRRV).* Human Interface Technology Lab report no:TR-95-1. Available from http://www.hitl.washington.edu/publications/tech-reports., 1995.

Sherman, William, R. & Craig, Alan, B. 'Literacy in Virtual Reality: a New Medium', *Computer Graphics,* **29** (4) (1995).

Vygotsky, L. *Mind in Society*. Harvard University Press, 1978.

Winn, W. *A Conceptual Basis for Educational Applications of VR.* Human Interface Technology Lab report no:TR-93-9 available from http://www.hitl.washington.edu/publications/tech-reports., 1993.

Index